THALLIUM IN THE ENVIRONMENT

Volume

29

in the Wiley Series in

Advances in Environmental
Science and Technology

JEROME O. NRIAGU, Series Editor

THALLIUM IN THE ENVIRONMENT

Edited by

Jerome O. Nriagu

Department of Environmental and Industrial Health
School of Public Health, The University of Michigan
Ann Arbor, Michigan

A WILEY-INTERSCIENCE PUBLICATION

JOHN WILEY & SONS, INC.

New York • Chichester • Weinheim • Brisbane • Singapore • Toronto

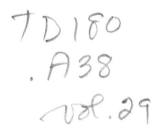

Library of Congress Cataloging-in-Publication Data:

Thallium in the environment / edited by Jerome O. Nriagu.
     p.   cm. — (Wiley series in advances in environmental science and technology ; v. 29)
    "A Wiley-Interscience publication."
    Includes index.
    ISBN 0-471-17755-5 (cloth : alk. paper)
    1. Thallium—Environmental aspects.   I. Nriagu, Jerome O.
II. Series: Advances in environmental science and technology ; v. 29.
TD196.T45T43   1998
363.738—dc21

Printed in the United States of America

10 9 8 7 6 5 4 3 2 1

CONTRIBUTORS

Dorothea Appenroth, Institute of Pharmacology and Toxicology, Friedrich Schiller University, D-07740 Jena, Germany

Cristina Ariño, Department of Analytical Chemistry, University of Barcelona, Av. Diagonal 647, 08028 Barcelona, Spain

Tom Brismar, Department of Clinical Neurophysiology, Karolinska Hospital, 17176 Stockholm, Sweden

Chiu L. Chou, Marine Environmental Sciences Division, Science Branch, Maritimes Region, Fisheries and Oceans Canada, P.O. Box 550, Halifax, Nova Scotia, Canada B3J 2S7

Ana del Peso, National Institute of Toxicology, P.O. Box 863, 41080 Sevilla, Spain

José Manuel Díaz-Cruz, Department of Analytical Chemistry, University of Barcelona, Av. Diagonal 647, 08028 Barcelona, Spain

Miquel Esteban, Department of Analytical Chemistry, University of Barcelona, Av. Diagonal 647, 08028 Barcelona, Spain

Elaine M. Faustman, Department of Environmental Health, University of Washington, 422 Roosevelt Way NE, 100, Seattle, Washington 98105-6099

Stephan Gambaryan, Institute of Evolutionary Physiology and Biochemistry of the Russian Academy of Sciences, St. Petersburg, Russia

Sonia Galván-Arzate, Departamento de Neuroquímica, Instituto Nacional de Neurología y Neurocirugía, Secretaria de Salud, Insurgentes sur 3877, México D.F., 14269 México

Cesarina Gregotti, Institute of Pharmacology, Medical School, University of Pavia, Piazza Botta 10, 27100 Pavia, Italy

Daniel I. Kaplan, Pacific Northwest National Laboratory, P.O. Box 999, Mailstop K6-81, Richland, Washington 99352

Tser-Sheng Lin, Department of Environmental and Industrial Health, School of Public Health, University of Michigan, Ann Arbor, Michigan 48109

Shas V. Mattigod, Pacific Northwest National Laboratory, P.O. Box 999, Mailstop K6-81, Richland, Washington 99352

John D. Moffatt, Marine Environmental Sciences Division, Science Branch, Maritimes Region, Fisheries and Oceans Canada, P.O. Box 550, Halifax, Nova Scotia, Canada B3J 2S7

v

JEROME O. NRIAGU, Department of Environmental and Industrial Health, School of Public Health, University of Michigan, Ann Arbor, Michigan 48109

GUILLERMO REPETTO, National Institute of Toxicology, P.O. Box 863, 41080 Sevilla, Spain

MANUEL REPETTO, National Institute of Toxicology, P.O. Box 863, 41080 Sevilla, Spain

CAMILO RÍOS, Departamento de Neuroquímica, Instituto Nacional de Neurología y Neurocirugía, Secretaria de Salud, Insurgentes sur 3877, México D.F., 14269 México

MANFRED SAGER, Bundesamt und Forschungszentrum für Landwirtschaft, Institut für Agrarökologie, Spargelfeldstraße 191, A-1226 Wien, Austria

HOMAYOUN TABANDEH, Moorfields Eye Hospital, London, England EC1V 2PD

BERNHARD W. VINK, Department of Geology, University of Botswana, Private Bag 0022, Gaborone, Botswana

CONTENTS

INTRODUCTION
TO THE SERIES

The deterioration of environmental quality, which began when mankind first congregated into villages, has existed as a serious problem since the industrial revolution. In the second half of the twentieth century, under the ever-increasing impacts of exponentially growing population and of industrializing society, environmental contamination of the air, water, soil, and food has become a threat to the continued existence of many plant and animal communities of various ecosystems and may ultimately threaten the very survival of the human race. Understandably, many scientific, industrial, and governmental communities have recently committed large resources of money and human power to the problems of environmental pollution and pollution abatement by effective control measures.

Advances in Environmental Science and Technology deals with creative reviews and critical assessments of all studies pertaining to the quality of the environment and to the technology of its conservation. The volumes published in the series are expected to service several objectives: (1) stimulate interdisciplinary cooperation and understanding among the environmental scientists; (2) provide the scientists with a periodic overview of environmental developments that are of general concern or that are of relevance to their own work or interests; (3) provide the graduate student with a critical assessment of past accomplishment, which may help stimulate him or her toward the career opportunities in this vital area; and (4) provide the research manager and the legislative or administrative official with an assured awareness of newly developing research work on the critical pollutants and with the background information important to their responsibility.

As the skills and techniques of many scientific disciplines are brought to bear on the fundamental and applied aspects of the environmental issues, there is a heightened need to draw together the numerous threads and to present a coherent picture of the various research endeavors. This need and the recent tremendous growth in the field of environmental studies have clearly made some editorial adjustments necessary. Apart from the changes in style and format, each future volume in the series will focus on one particular theme or timely topic, starting with Volume 12. The author(s) of each pertinent section will be expected to critically review the literature and the most important recent developments in the particular field; to critically evaluate new concepts, methods, and data; and to focus attention on important unresolved or controversial questions and on probable future trends. Monographs embodying the results of unusually extensive and well-rounded investigations will also be published in the series. The net result of the new editorial policy should be more integrative and comprehensive volumes on key environmental issues and pollutants. Indeed, the development of realistic standards of environmental quality for many pollutants often entails such a holistic treatment.

JEROME O. NRIAGU, Series Editor

PREFACE

Thallium is more toxic to mammals than mercury, cadmium, lead, copper or zinc, and has been responsible for many accidental, occupational, deliberate, and therapeutic poisonings since its discovery in 1861. Its past medical uses included treatment for gonorrhea, syphilis, dysentery, tuberculosis and ring worm of the scalp. It was also extensively employed as a rat poison (now outlawed in the United States but not in most developing countries). The semiconductor industry is the main user of (metallic) thallium, but substantial quantities also go into the production of cardiac imaging, highly refractive optical glass, and some alloys. Small amounts are used in superconductivity research, radiation detection equipment and photosensitive devices. Thallium compounds have a wide range of industrial uses. Thallium chloride is used as a catalyst in chlorination. Because of its high specific gravity, the solution of T1 acetate is used in separating ore constituents by flotation. Thallium nitrate is used with other compounds as signals at sea and in the production of fireworks, photocells and as an oxidizing agent in organic synthesis. The oxide is used in the manufacture of highly refractory glass and in the production of artificial gems. Thallium sulfate is used in the semiconductor industry, in low range thermometers, optical systems, and photoelectric cells. Since thallium salts are generally volatile, the incineration of any solid wastes containing T1 products can be an important source of this element in the atmosphere. The increasing use of thallium in emerging technologies has raised new concerns about health risks and environmental toxicology of this element.

The central and peripheral nervous systems are the target organs most affected by chronic thallium poisoning. The effects on the reproductive system include changes in reproductive behavior, degeneration of the male and female genital tracts, and abnormalities in egg and fetal development and survival. Growth retardation, beak deformity, microphthalmia, microcephaly, reduced size of all skeletal elements, as well as micromelia, achodroplasia and abnormal gonad development have also been associated with chronic thallium poisoning. Papillomas, precancerous lesions and cancers in the genital tracts of mice exposed to thallium have also been noted. Symptoms similar to those of thallium poisoning have been noted in many wildlife populations in the Great Lakes basin, and one report claimed that 9 out of the 34 bald eagles found sick or dying in 1971–1972 in the United States were poisoned by thallium.

One objective of this volume is to raise the public consciousness about the potential risks of thallium in our environment.

In its chemical behavior and biological effects, thallium resembles the heavy metals (lead, gold, and silver) on the one hand and the alkali metals on the other. At the molecular level, thallous ion inhibits the movement of K^+ across the mitochondrial membrane as well as the Na^+/K^+ ATPase pump, and hence affects the oxidative phosphorylation and other energy-producing reactions. $Tl(I)$ binds more strongly than K^+ at specific sites and thus tends to act as a cumulative poison. Monovalent thallium also has a strong affinity for the sulfhydryl groups and can inhibit many enzyme reactions such as succinic dehydrogenase, protease and monoamine oxidase. The combination of highly divergent properties in one element has led some authors to refer to thallium as an enigmatic or paradoxical metal. The differing biochemistry of its principal redox forms also implies that thallium can exercise a multi-target organ toxicity on many organisms. The fascinating chemistry and high toxicity potential make thallium and its compounds of particular scientific interest and environmental concern.

This volume represents an attempt to bring together the global literature on biological, chemical, and clinical studies on thallium in various environmental media. Individual chapters cover the sources, cycling, speciation, fate as well as human and ecosystem effects. The chapters are written by leading experts in their fields and the transdisciplinary volume provides an integrated account of current knowledge on one of the most enigmatic metals known to mankind. The authors have been asked to focus on general principles of thallium behavior rather than on systematic compilation of published data. The volume should thus be of interest to graduate students and practicing scientists in the fields of environmental science and engineering, toxicology, public health, and environmental control. More importantly, it is addressed to everyone who is concerned about the impact of metallic pollutants on our health and our life support system.

JEROME O. NRIAGU

Ann Arbor, Michigan

ADVANCES IN ENVIRONMENTAL SCIENCE AND TECHNOLOGY
Jerome O. Nriagu, Series Editor

THALLIUM IN THE ENVIRONMENT

1

HISTORY, PRODUCTION, AND USES OF THALLIUM

Jerome O. Nriagu

*Department of Environmental and Industrial Health, School of
Public Health, University of Michigan, Ann Arbor, MI 48109*

1.1. INTRODUCTION

Thallium is like several elements in one in terms of its chemical properties,
natural occurrence, geochemical behavior, and biological effects. Attempts
have been made to relate the eccentricity of the element to the properties
and behavior of its atom (Table 1) and to the coordination and structural
chemistries of the compounds (Downes, 1993); such a discussion is outside
the objectives of this chapter. Thallous salts resemble those of alkali metals
in many respects—the ionic mobilities are closely similar, the hydroxides are
soluble and strong bases, the phosphates are soluble, the azides are not

Thallium in the Environment, Edited by Jerome O. Nriagu.
ISBN 0-471-17755-5. © 1998 John Wiley & Sons, Inc.

Table 1   Physical and Chemical Properties of Thallium

Property	Value
Atomic number	81
Ground-state electronic configuration	$[\text{Xe}]4f^{14}5d^{10}6s^26p^1$
Spin-own-orbit coupling constant (A/cm)	5195
Ionization energies (kJ/mol)	
$\quad$ M $\rightarrow$ M$^+$	589.3
$\quad$ M$^+$ $\rightarrow$ M^{2+}	1971.0
$\quad$ M^{2+} $\rightarrow$ M^{3+}	2878
$\quad$ M^{3+} $\rightarrow$ M^{4+}	~4900
Electronic affinity [Tl(g) $\rightarrow$ Tl$^-$(g) (kJ/mol)]	~20
Effective nuclear charge, Z	
$\quad$ Slater value	5.00
$\quad$ Clementi value	12.25
$\quad$ Froese–Fischer value	13.50
Melting point (K)	577
Boiling point (K)	1746
Density (g/cc)	11.85
Electrical resistivity (Ω m)	18×10^{-8}
Thermal conductivity at 300 K (W/m/K)	46.1
ΔH_{vap} at 298 K (kJ/mol)	180.9
$\Delta H_{\text{atomization}}$ at 298 K (kJ/mol)	182.2
Standard entropy, S at 298 K (kJ/mol)	64.18
Electronegativity	
$\quad$ Pauling scale	1.62 (Tl$^+$); 2.04 (Tl^{3+})
$\quad$ Allred scale	1.44
$\quad$ Sanderson scale	1.96 (0.99, Tl$^+$; 2.25, Tl^{3+})
$\quad$ Pearson scale	3.2
Atomic (metallic) radius (Å)	1.704 (α-form)
Single-bond covalent radius (Å)	1.55
van der Waals radius (Å)	2.00
Ionic radius for sixfold coordination (Å)	
$\quad$ Tl$^+$	1.50
$\quad$ Tl^{3+}	0.885
ΔH_f° [Tl$^+$(g)] (kJ/mol)	777.7
ΔH_f° [Tl^{3+}(g)] (kJ/mol)	5639.2
ΔG_f° [Tl^{3+}(aq)] (kJ/mol)	+214.6
ΔH_f° [Tl^{3+}(aq)] (kJ/mol)	+196.6
Standard redox potentials (V)	
$\quad$ Tl$^+$ + e^- $\rightarrow$ Tl(s)	−0.336
$\quad$ Tl^{3+} + $3e^-$ $\rightarrow$ Tl(s)	+0.741
$\quad$ Tl^{3+} + $2e^-$ $\rightarrow$ Tl$^+$	+1.28

explosive, the metallic ions do not form strong complexes, the metals form stable polysulfides, and the salts are often isomorphous (Durrant and Durrant, 1970). Silver salts likewise are analogous to those of monovalent thallium in that they are colorless and mostly anhydrous; the solubilities of its halides, sulfide and chromate, are low; and the melting points of the halides are very similar. Because of similarity ion ionic size, Tl(I) can replace Rb, K, Na, Pb, Ag, Au, and Zn in some minerals and biogeochemical systems. On the other hand, Tl(III) behaves very much like other trivalent cations in its acceptor properties and the insolubility of its oxyhydroxide compounds. Thallium thus displays a wide diversity in its biogeochemical behavior and in consequence is very widely dispersed in nature in low concentrations and has been recovered from sulfide minerals deposited in widely differing environments spanning several geologic eras. Thallium's infinite variety of chemical personality is also reflected in the wide range of potential commercial applications, although its uses are tempered by high cost and toxicity of the element. The nonconforming behavior provides a mine field of research opportunity for chemists and presents a challenge to efforts to turn this enigmatic element into useful commercial products.

1.2.   HISTORY

Thallium owes its discovery and even its name to developments in spectroscopic analysis of light emitted by action of flame or electric spark on metal compounds. Primacy in actual discovery of thallium is mired in national intrigue between the British and French scientist (Weeks, 1968). First evidence for the existence of the element, thallium, was obtained in 1861 by Sir William Crookes (1832–1919). While doing a spectroscopic examination of seleniferous residues from a sulfuric acid factory in Tilkerode in the Harz Mountains, Crookes ran across a beautiful green line that had never been seen before. He concluded that the material must contain a new element, which he named thallium, "from the Greek θαλλσφ or Latin *thallus,* a budding twig—a word which is frequently employed to express the beautiful green tint of young vegetation; and which I have chosen because the green line which it communicates to the spectrum recalls with peculiar vividness the fresh color of vegetation in the spring." He announced the results in a paper entitled "On the existence of a new element, probably of the sulfur group" in the March 30, 1861, issue of *Chemical News,* a journal he founded in 1859 and edited until 1906. Sir Crookes did additional studies on his discovery and in 1862 was awarded a prize at an international exposition for a specimen labeled "Thallium, a new metallic element."

Claude-Auguste Lamy (1820–1878) independently observed the green line of thallium in March 1862 in a sample of selenium extracted from slime of sulfuric acid plant at Loos where Belgian pyrites were roasted. On June 23,

1862, he presented a 14-g ingot of metallic thallium to the Academie des Sciences and detailed his recovery procedure as follows:

> When these deposits [from roasting of pyrites] are heated almost dry, with approximately an equal volume of aqua regia until the acid almost disappears, and the mass is then taken up with twice its weight of boiling water, one sees formed in the liquid as it cools an abundance of yellow crystalline plates which, when purified by several successive recrystallizations, give a magnificent compound of thallium sesquichloride. When this chloride is submitted to the decomposition action of the electric current from four or five Bunsen cells, for example, there appears at the negative pole pure thallium. This is the experiment by which we have, for the first time, isolated the new metal. (Lamy, 1862; translation from Weeks, 1968, p. 611)

From careful experimental studies, Lamy showed that the new element formed two series of salts depicting monovalent (thallous) and trivalent (thallic) oxidation states. Reports of the new metal quickly aroused the interest of other chemists. By the end of 1863, about 28 papers had been published covering different aspects of thallium, including the synthesis and properties of organic and inorganic compounds, toxicological effects, and occurrence in minerals, seawater, and vegetation (Doan, 1899). After reviewing the volume of information from these early studies, Jean Baptiste-Andre Dumas proclaimed that "thallium is destined to be epoch-making in the history of chemistry by the remarkable contrast between its chemical nature and its physical properties. It is no exaggeration to say that from the point of view of classification generally accepted for metals, thallium offers a combination of contradictory properties which would entitle one to call it the paradoxical metal, the ornithorynchus of the metals" (Pelouze et al., 1867; translation in Weeks, 1968, p. 612). By 1870, over 100 articles had been published and the basic chemistry of thallium had become defined (Doan, 1899).

It was, initially, difficult to place the new element in the periodic classification of the elements. The physical properties of thallium including specific gravity, hardness, appearance, melting point, and electrical conductivity were found to be similar to those of lead. The chemical properties of thallous salts, such as sparing solubility, color of the halides and chromate, and the behavior of their aqueous solutions toward lead, were also found to closely resemble those of lead. The differences in valency and other characteristics, however, were too pronounced to exclude it from the lead family of elements. On the other hand, Tl(I) resembles the alkali metals in the flame spectra, solubility of the hydroxide, sulfate, and carbonate in water, ready oxidation of the metal in air, existence of thallium alums, and the isomorphism of its salts with those of potassium, cesium, and rubidium. The isomorphism of the sulfates, alums, phosphates, tartrates, dithionates, perchlorates, and chlorates were, however, too divergent to allow the inclusion of thallium in the family of alkali metals.

Trivalent thallium had analogies with the aluminum group of elements in many respects. In his periodic classification of the elements, D. I. Mendeleev argued strongly in favor of placing thallium in the aluminum family. He noted that in placing the elements in the order of their atomic weights, aluminum comes between the magnesium and silicon and thallium between mercury and lead. He emphasized the analogy between the higher oxides of mercury, lead, and thallium with those of magnesium oxide, silica, and alumina, respectively. Magnesium and mercuric oxides are strongly basic, aluminum and thallic oxides are feebly basic, while lead and silicon oxides are feebly acidic. Mendeleev inferred that HgO, Tl_2O_3, and PbO_2 are peroxide relatives to the lower oxides Hg_2O, Tl_2O, and PbO and give off oxygen when heated (Mellor, 1956).

The toxicity of thallium and its compounds became immediately apparent and was discussed by W. Crookes in 1863, A. Lamy in 1863, J. J. Pualet in 1863, L. Grandeau in 1863, and B. Stadion in 1868 (Mellor, 1956). In spite of these early studies, thallium was introduced therapeutically for syphilis in 1883, for night sweats of tuberculosis in 1898, for depilation in 1897, for ringworm of the scalp in 1919, and as a pesticide in 1920. In the late 1920s, Koremlu Cream for removing unwanted hair entered the United States market and became the source of thallium poisoning for a lot of people. There is a long trail of evidence to show that the commercialization of thallium in consumer products was attended by numerous cases of poisonings (see Prick et al., 1955). By 1934, 778 cases of thallium poisoning, including 46 fatalities, had been reported in the scientific literature, and a 1947 review of the medical literature dealing with thallium toxicology uncovered over 300 references (Beliles, 1991).

In 1866, during an analysis of selenium minerals (eucairite and berselianite) from Skrikerum in Sweden, Baron Nils Adolf Erik Nordenskiold detected significant concentrations of thallium. Subsequent examinations led to the discovery of thallium selenide [$(Cu, Tl, Ag)_2Se$], a new mineral that he named *cookesite*, in honor of Sir William Cookes, the discoverer of the element (Weeks, 1968). Although many thallium-containing minerals have been discovered and described (see Table 2), the following are the most common (see Palache et al., 1949):

Lorandite, $TlAsS_2$, discovered by J. A. Krenner in 1894 comingled with realgar (HgS) at Allchar, Greece, and named after the physicist Eotvos Lorand of Budapest

Hutchisonite, $(Tl,Cu,Ag)_2S.PbS.2As_2S_3$, discovered in 1905 by R. H. Soby and G. F. H. Smith in dolomite rocks at Lengenbach in the Binnental, Valais, Switzerland, and named after Dr. Arthur Hutchinson (1866–1937), former professor of mineralogy in the University of Cambridge

Urbaite, $TlAs_2SbS_5$, discovered in 1914 in Allchar, Greece, by B. Jezek and F. Krehlik and named after Karl Vrba (1845–1922), a Bohemian mineralogist

Table 2   Thallium Minerals

Sulfides, Antimonides, and Selenides

Crookesite, $(Cu,Tl,Ag)_2Se$
Sabatierite, Cu_6TlSe_4
Cuprostibite, $Cu_2(Sb, Tl)$
Carlinite, Tl_2S
Picotpaulite, $TlFe_2S_3$
Raguinite, $TlFeS_2$
Thalcusite, $Tl_2Cu_3FeS_4$
Bukovite, $Tl_2(Cu,Fe)_4Se_4$
Rayite, $Pb_8(Ag,Tl)_2Sb_8S_{21}$
Rathite, $(PbTl)_3As_5S_{10}$
Hatchite, $(Pb,Tl)_2AgAs_2S_5$
Thalfenisite, $Tl_6(Fe,Ni,Cu)_{25}S_{26}Cl$
Lorandite, $TlAsS_2$
Ellisite, Tl_3AsS_3
Imhofite, $Tl_6CuAs_{16}S_{40}$
Simonite, $TlHgAs_3S_6$
Christite, $TlHgAsS_3$
Routhierite, $TlHgAsS_3$
Hutchinsonite, $(Tl,Pb)_2As_5S_9$
Wallisite, $(Cu,Ag)TlPbAs_2S_5$
Bernardite, $Tl(As,Sb)_5S_8$
Vrbaite, $Tl_4Hg_3Sb_2As_8S_{20}$
Rebulite, $Tl_5Sb_5As_8S_{22}$
Gillulyite, $Tl_3(As,Sb)_8S_{13}$
Weissbergite, $TlSbS_2$
Rohaite, Cu_5TlSbS_2
Vaughanite, $TlHgSb_4S_7$
Chalcothallite, $Tl_2(Cu,Fe)_6SbS_4$
Pierrotite, $Tl_2Sb_6As_4S_{16}$
Parapierrotite, $Tl(Sb,As)_5S_8$
Criddleite, $TlAg_2Au_3Sb_{10}S_{10}$
Chabournéite, $(Tl,Pb)_5(Sb,As)_{21}S_{34}$

Oxides and Sulfates

Avicennite, Tl_2O_3
Monsmedite, $K_2O \cdot Tl_2O_3 \cdot 8SO_3 \cdot 15H_2O$

1.3.   OCCURRENCE

Thallium is widely distributed in the natural environment, but the concentrations are generally very low. It occurs naturally as two stable isotopes: ^{203}Tl and ^{205}Tl with relative abundances of 29.5 and 70.5%, respectively (Sahl et al., 1978). The ^{205}Tl/^{203}Tl isotopic ratio of 2.378 does vary with geological processes (Ostic, 1967) and hence cannot be used as a fingerprint for sources of Tl in the environment. The isotopic compositions of terrestrial and meteoritic Tl agree within 1% (Ostic, 1967), implying that there is no excess radiogenic ^{205}Tl in meteorites.

The following radioactive Tl isotopes can be obtained by irradiating ^{203}Tl and ^{205}Tl with thermal neutrons:

Radionuclide	Half-life	Radionuclide	Half-life
^{201}Tl	3.038 days	^{202}Tl	12.23 days
^{207}Tl	4.77 min	^{206}Tl	3.76 min
^{209}Tl	2.2 min	^{208}Tl	3.053 min
		^{210}Tl	1.30 min

Because of its longer half-life, ^{204}Tl is often employed in radiometric studies.

Atomic abundance of thallium in the solar system (normalized to Si = 10^6 atoms) is estimated to be 0.182 (Sahl et al., 1978). The Tl content of Earth is 0.004 μg/g (Brooks and Ahrens, 1961), compared to an abundance in Earth's crust of 0.8 μg/g (Sahl et al., 1978). A 200-fold enrichment in the crust compared to the mantle and core is suggested. Average Tl contents of five types A and B lunar samples and eight types C and D samples collected by Apollo 11 astronauts were 0.63 and 2.44 ng/g, respectively (Keays et al., 1970).

Average concentrations of thallium in major rock formations are shown in Table 3. According to its distribution in igneous rocks, thallium is a lithiophile (occurs with or in silicates) element. This is not surprising since its geochemistry closely follows that of potassium and rubidium. The K/Tl and K/Rb ratios in igneous rocks are in the ranges of 4,000–112,000 and 19–590, with the averages being 30,000 and 150, respectively (Sahl et al., 1978). Thallium contents of potassium feldspar generally fall in the range of 0.5–50 μg/g although values as high as 600 μg/g have been reported in potassium feldspars from strongly differentiated pegmatites (see Sahl et al., 1978). Thallium also forms sparingly soluble sulfides and is a minor constituent of pyrites and other metal sulfides and thus shows some chalcophile characteristics.

Thallium is generally enriched in fine-grained sediments (especially organic- and sulfide-rich clays and silts) compared to sandstones and carbonates (Sahl et al., 1978). Average concentration of Tl in soils is estimated to be 0.2 μg/g, the range being from <0.01 to over 5 μg/g in mineralized areas (IPCS, 1996). Most of the reported Tl levels in coals are questionable; the average concentration is provisionally estimated to be 0.05 μg/g.

Thallium minerals are very rare in nature and their deposits are so small as to have no commercial significance. Most of the thallium minerals so far identified are sulfides (Clark, 1993). The compositions of some of the thallium minerals are listed in Table 2.

Table 3   Natural Abundance of Thallium in Geological Material

Rock Type	Tl Conc. (μg/g)	Reference
Igneous Rocks		
Ultramafic	0.05	Sahl et al., 1978
Mafic	0.18	Sahl et al., 1978
Intermediate	0.55	Sahl et al., 1978
Granitic-rhyolitic	1.7	Sahl et al., 1978
Alkalic	1.2	Sahl et al., 1978
Sedimentary Rocks		
Shales	0.7	Shaw, 1957
Sandstones	0.03[a]	
Carbonates	0.05	Sahl et al., 1978
Greywackes	0.3	Sahl et al., 1978
Deep-sea sediments:		
Carbonates	0.16	Sahl et al., 1978
Clays	0.6	Sahl et al., 1978
Manganese nodules	100	Sahl et al., 1978
Freshwater sediments	0.35	IPCS, 1996
Metamorphic Rocks		
Eclogites	0.3[a]	
Schist	0.6[a]	
Gneiss	0.37[a]	
Slates and phyllites	0.46[a]	
Quartzose	0.02[a]	
Coal	0.05[a]	
Soil	0.2[a]	
Terrestrial plants	0.015[a]	
Land mammals	0.005[a]	

[a] Compiled from the literature.

1.4. PRODUCTION

Although a number of thallium minerals occur in nature (see above), these have no commercial importance as a source of this metal. It occurs, however, in relatively high concentrations in some sulfide deposits. Because of the high volatility of its compounds, thallium is enriched in dust particles from the smelting of such ores. In recent times, thallium is recovered primarily as a minor by-product of zinc smelting. Many zinc ores contain >2 μg/g of thallium, which become concentrated in flue dusts and residues, and these provide the commercial source for thallium production. Minor amounts of thallium have also been recovered from lead, copper, and iron smelters and from sulfuric acid plants. The generation of Tl-containing residues is often higher than the Tl requirements, and the unprocessed portion can become a bothersome stock in the industry.

Recovery of Tl from zinc industry residue involves a multistep process designed to take advantage of the difference in solubilities of thallium compounds and those of other metal impurities. Thallium sulfide is insoluble in alkaline solution but soluble in acid, making it possible to separate thallium from Group I elements. Because thallium sulfate is soluble in water, it can be removed from lead sulfate. The low solubility of thallium chloride in cold water is used in separation of thallium from cadmium, copper, tellurium, and zinc. The industrial process used depends on the relative proportions of different minerals and thus varies from smelter to smelter. In one method that yields thallium as a by-product of cadmium recovery, the flue dusts are leached with sulfuric acid to form cadmium and thallium sulfates. After removing the impurities by precipitation, the solution is electrolyzed to yield a Cd–Tl alloy containing 5–50% thallium. After converting the alloy to hydroxides by treatment with hot water and steam, sodium carbonate is added to precipitate the cadmium. The filtrate is treated with hydrogen sulfide to precipitate thallium sulfide, which is then dissolved in sulfuric acid. The thallium sulfate is either (a) electrolyzed to recover pure metallic thallium sponge or (b) the thallium is precipitated on zinc plates that are then heated to form thallium sponge. The sponge is compressed or fused into 6.5-oz sticks, 1-lb ingots, 34-lb ingots or wires with purity in the range of 99.90–99.999% (Bascle, 1980).

Industrial demand for thallium is low and nowadays limited to high-tech industries. World production of Tl has remained fairly constant during the 1970s and 1980s and has declined in the 1990s as a result of the growing concern about its environmental toxicity. Worldwide production figures are estimated to be 12 tons in 1975, 17 tons in 1979, 18 tons in 1985, 17 tons in 1987–88. The geographic distribution of thallium production should be similar to that of zinc mine production. In many developing countries, however, the thallium is not recovered but discharged into the environment. Countries that have produced commercial quantities of thallium include Germany, Belgium, The Netherlands, the former Soviet Union, Mexico, Canada, United States, and Japan (Bascle, 1980). Production was discontinued in the United States in 1981. In terms of supply–demand relationships, it is clear that if the need arises, the production of thallium can be increased sharply without any difficulty.

Based on average thallium content of 2.2 $\mu g/g$ in zinc sulfides, the world's reserves of thallium are estimated to be about 380 tons (Llewellyn, 1991). In addition, about 350 tons of thallium are in coal ash (Bascle, 1980), and about 150 tons are in pyrites and other metal sulfides and selenides. World resources of thallium thus total about 880 tons.

1.5. USES

Thallium remained an industrial curiosity for over half a century after it was first discovered; poor mechanical properties make pure thallium an undesirable industrial material (Korenman, 1969). Expanded commercial application began with the patenting of a photosensitive cell using thallium oxysulfide in 1919. The cumulative index of *Chemical Abstracts* of 1995 lists over 150 uses and potential applications for thallium and its compounds. The most innovative applications are in the high-technology and future technology fields. In its use, thallium thus has a short history but what appears to be a rosy future.

The use of thallium salts as poisons for rodents and later as insecticides began in 1920 and for the next 45 years remained the principal use for this element. In 1965, the U.S. government prohibited the use of thallium salts as rodenticide because of the many reported cases of thallotoxicosis, and in 1973 the U.S. Environmental Protection Agency requested all retail outlets to surrender stocks of pesticides that still contained thallium (Bascle, 1980). Although thallium is no longer sold as pesticides in the developed countries, it is still used in many developing countries because of its cheapness. Medicinal uses for thallium have been discontinued in most countries. Throughout its history, thallium has been an instrument of murder or suicide.

Commercial applications for thallium metal or for alloys in which thallium is the principal component are very limited. Thallium, however, is a minor component of many useful alloys, although the growing concern for its toxicity is resulting in the phase-out of thallium in a number of these alloys. The semiconductor industry accounts for 60–70% of current thallium consumption by industries in the United States, the remainder going to the manufacture of highly refractive optical glass and cardiac imaging (Table 4; ATSDR, 1992). Important uses for thallium-containing alloys include:

1. Bearing alloys with various amounts of Pb (1–34%), Sb (1–10%), Tl (1–34%), and Cu (65–95%). Bearing alloys of silver (92–98%) and thallium (2–8%) have been used in situations requiring acid resistance

Table 4 Demand Pattern for Thallium in the United States

Year	Demand by Industrial Sectors (in kg/yr)				
	Electrical	Agricultural	Pharmaceutical	Other	Total
1969	1600	2450	150	350	4550
1970	1650	1000	100	350	3100
1971	700	1000	75	200	1950
1972	550	450	50	200	1250
1973	500	75	75	50	700
1974	700	75	75	150	1000
1975	500	0	100	300	900
1976	700	0	100	350	1250
1977	950	0	75	370	1400
1978	750	0	50	300	1100
1979	1100	0	75	425	1600

and low coefficient of friction. Alloys of gold (25–50%), silver (25–50%), and thallium (1–40%) also have high corrosion resistance, low coefficient of friction, and high fatigue (or endurance) limit.

2. Acid-resistant alloys such as 10% Tl, 20% Sn, 70% Pb and 20% Tl, 80% Pb have been used for mineral acid containers.

3. Corrosion-resistant alloys with 20–50% Tl, 50–80% Pb were used for electrolytic coating of iron, steel, and brass.

4. Antifriction alloys such as 8% Tl, 5% Sn, 15% Sb, 72% Pb.

5. Low-melting-point glasses containing some S, As, and Se. These glasses are fluid at 125–150°C and are more durable and less water soluble than ordinary glasses; changes in refractive index and density also confer improved optical characteristics on these glasses.

6. Electrode alloys such as 70% Pb, 20%, 10% Sn used as anodes for electrolytic deposition of copper.

7. Low-freezing-point alloys such as 81.5% Hg, 8.5% Tl, which has a freezing point of $-60°C$, are particularly suited for thermometers, switches, and other instruments employed in polar regions.

Thallium compounds have important commercial applications. Since World War II, large quantities of thallium halogenides (TlBr–TlI), sulfides (Tl$_3$VS$_4$; Tl$_3$NbS$_4$), and selenides (Tl$_3$PSe$_4$) have gone into military research and production of laser and acusto-optic equipment. Thallium halides transmit infrared wavelength light and are used in equipment for detection and signaling where visible radiation must be excluded. Crystals of thallium bromoiodide have been used as lenses, plates, and prisms in optical spectroscopic systems. Thallium oxysulfide is used to detect infrared radiation. Alkali halide crystals doped with thallium, such as sodium iodide activated with 0.1–0.2% thallium, are used in discriminator circuits of scintillator counters for measurements of α-particles, β-particles, and γ-rays, and for high-resolution imaging of soft X-rays. Cesium iodide activated with thallium has been used in scintillators for X-ray tomography.

Monovalent thallium carbonate, bromide, and iodide have been used in the manufacture of fiber (optical) glass with high refractive index and in the preparation of opaque black or brown glass. Thallium-impregnated glass is used as protective coatings on semiconductors, capacitors, and other electronic devices. Electrical conductivity of partially oxidized thallium sulfide changes upon exposure to light and is the basis of photocells with a maximum sensitivity at 900 μm; some thallium halides are also photoelectrically sensitive.

Various thallium compounds have been explored as catalysts in the oxidation, reduction, aromatization, esterification, and polymerization of organic molecules. Thallium oxide, for example, has been used as a catalyst in the synthesis of ammonia, in the chlorination of hydrocarbons, in reduction of nitrobenzene by hydrogen, in photopolymerization of methyl methacrylate, in the oxidation of hydrocarbons and olefins, and in epoxidation reactions

(Korenman, 1969; Smith and Carson, 1977). Minor amounts of various thallium compounds are used in analytical chemistry and chemical and biochemical research. Other uses for thallium compounds include fireworks (thallium nitrate), pigments (thallium chromate) and dyes, production of imitation jewels (thallium oxide), impregnation of wood and leather against fungi and bacteria, and in mineralogical separation (IPCS, 1996). Thiolate complexes of thallium have been used to study the environmental and toxicological impacts of metal biomethylation, molecular precursors to solid-state compounds of technological interest, and to explore the stereochemical activity of s^2 lone pair of electrons (Bosch et al., 1996). Thallium ions show excellent nuclear magnetic resonance (NMR) properties and have been used as a probe to emulate the biological functions of alkali metal ions, especially K^+ and Na^+ (Downes, 1993).

Addition of monovalent thallium halides to mercury lamps increases the intensity of the light and its spectrum. Flames colored with thallium were once used as a source of monochromatic light (Mellor, 1956). Discharge lamps filled with mixtures containing thallium have been used for black-body chromaticity.

Thallium radionuclides have a number of important applications. Radiogenic ^{205}Tl has been employed in measuring time, ^{204}Tl for determining the thickness of material, and ^{201}Tl in scintillography imaging of the heart, liver, thyroid, and testes and in the diagnosis of melanoma.

A major development in potential industrial application for thallium occurred with the recent achievement of superconductivity in ceramic compounds of the general formula $(M^IO)_m X_2^{II} Ca_{n-1} Cu_n O_{2n+2}$ where M^I can be Tl, Bi, Pb or a mixture thereof, $m = 1$ or 2 (2 only if M^I is Bi), and M^{II} is Ba or Sr (Sheng and Hermann, 1988; Kahwa et al., 1992). A number of thallium-containing superconducting material and their critical transition temperatures for zero resistivity (T_c) are shown in Table 5. Thallium-containing systems

Table 5   Superconductivity Transition Temperatures, T_c, for Thallium-Containing Compounds[a]

General Formula	Specific Compound	T_c (K)
$TlCa_{1-n}Ba_2Cu_nO_{3+2n}$	$TlBa_2Cu_nO_5$	~60
	$TlCaBa_2Cu_2O_7$	85
	$TlCa_2Ba_2Cu_3O_9$	110
	$TlCa_3Ba_2Cu_4O_{11}$	122
$Tl_2Ca_{n-1}Ba_2Cu_nO_{4+2n}$	$Tl_2Ba_2CuO_6$	81
	$Tl_2CaBa_2Cu_2O_8$	110
	$Tl_2Ca_2Ba_2Cu_3O_{10}$	125
	$Tl_2Ca_3Ba_2Cu_4O_{10}$	119
$(Tl,Pb,Bi)Ca_{1-n}Sr_2Cu_nO_{3+2n}$	$(Tl,Pb)Ca_2Sr_2Cu_3O_9$	122
	$(Tl,Bi)CaSr_2Cu_2O_7$	90
	$(Tl,Pb)CaSr_2Cu_2O_7$	90
	$(Tl,Bi)Sr_2CuO_5$	50

[a] From Evans (1993) and Polmear (1993).

have advantages over the more popular $YBaCu_3O_7$, such as higher T_c, ease of preparation, and being cooled with liquid nitrogen instead of more expensive helium (Polmear, 1993; Evans, 1993). The superconducting oxides hold a promise for new and exciting technological developments. The high toxicity of thallium, however, must be seen as a dark cloud surrounding future large-scale commercialization of these material.

REFERENCES

ATSDR (1992). *Thallium.* Agency for Toxic Substances and Disease Registry, U.S. Public Health Service, Department of Health and Human Services, Research Triangle Park, NC.

Bascle, R. J. (1980). Thallium. In *Mineral Facts and Problems.* Bureau of Mines Bulletin 671, U.S. Department of the Interior, Washington, DC, pp. 931–936.

Beliles, R. P. (1991). Thallium. In *Patty's Industrial Hygiene and Toxicology,* 4th ed., Clayton, G. D., and Clayton, F. E. (eds.). Wiley, New York, Vol. II, Part C, pp. 2235–2248.

Bosch, B. E., Eisenhawer, M., Kersting, B., Kirschbaum, K., Krebs, B., and D. M. Glolando (1996). Synthesis, properties and crystal structures of benzene-1,2-dithiolato complexes of thallium(I) and -(III). *Inorg. Chem.,* **35,** 6599–6605.

Clark, A. M. (1993). Hey's Mineral Index. Chapman and Hall, London

Crookes, W. (1861). On the existence of a new element, probably of the sulfur group. *Chem. News,* **3,** 193–195.

Doan, M. (1899). *Index to the Literature on Thallium, 1861–1896.* Smithsonian Institution, Washington, DC.

Downes, A. J. (1993). Chemistry of Group 13 metals: Some themes and variations. In *Chemistry of Aluminium, Gallium, Indium and Thallium,* Downes, A. J. (ed.), Blackie Academic & Professional, London, pp. 1–80.

Durrant, P. J., and Durrant, B., (1970). *Introduction to Advanced Inorganic Chemistry.* Wiley, New York, pp. 590–600.

Evans, K. A. (1993). Properties and uses of oxides and hydroxides. In *Chemistry of Aluminium, Gallium, Indium and Thallium,* Downes, A. J. (ed.), Blackie Academic & Professional, London, pp. 248–291.

IPCS (1996). *Thallium.* International Program on Chemical Safety, Environmental Health Criteria 182, World Health Organization, Geneva, Switzerland.

Kahwa, I. A., Miller, D., Mitchell, M., Fronczek, F. R., Goodrich, R. G., Williams, D. J., O'Mahoney, C. A., Slawin, A. M. Z., Ley, S. V., and Groombridge, C. J. (1992). *Inorg. Chem.,* **31,** 3963–3978.

Keays, R. R., Granapathy, R., Laul, J. C., Anders, E., Herzog, G. F., and Jeffery, P. M. (1970). Trace elements and radioactivity in lunar rocks: Implications for meteorite in fall, solar wind flux, and formation conditions of the moon. *Science,* **167,** 490–493.

Korenman, I. M. (1969). *Thallium.* Ann Arbor-Humphrey Science, Ann Arbor, MI.

Lamy, C. A. (1862). De l'existence d'un nouveau metal, le thallium. *Compt. Rend.,* **54,** 1255–1258.

Llewellyn, T. O. (1991). Thallium. In *Minerals Yearbook.* Bureau of Mines, U.S. Department of the Interior, Washington, DC.

Mellor, J. W. (1956). *A Comprehensive Treatise on Inorganic and Theoretical Chemistry,* Vol. V. Longans, Green, London, pp. 406–479.

Ostic, R. G. (1967). *Isotopic Composition of Thallium in Meteorites.* U.S. Atomic Energy Commission Report NYO-844-71, Washington, DC.

Palache, C., Berman, H., and Frondel, C. (1949). *The System of Mineralogy.* Wiley, New York.

Pelouze, T. J., Deville, H. S. and Dumas, J. B. A. (1862). Rapport sur un Memoire de M. Lamy, relatif au thallium. *Compt. Rend.,* **55,** 836–838.

Polmear, I. J. (1993). The elements. In *Chemistry of Aluminium, Gallium, Indium, and Thallium,* Downes, A. J. (ed.), Blackie Academic & Professional, London, pp. 81–110.

Prick, J. J. G., Smitt, W. G. S., and Muller, L. (1955). *Thallium Poisoning.* Elsevier, Amsterdam.

Sahl, K., Albuquerque, C. A. R., and Shaw, D. M. (1978). Thallium. In *Handbook of Geochemistry,* Wedepohl, K. H. (ed.). Springer-Verlag, Berlin, Vol. II, Section 5.

Shaw, D. M. (1957). The geochemistry of gallium, indium and thallium—a review. *Phys. Chem. Earth,* **2,** 164–178.

Sheng, Z. Z., and Hermann, A. M. (1988). *Nature,* **332,** 55–57.

Smith, I. C., and Carson, B. L. (1977). *Trace Metals in the Environment: Volume 1, Thallium.* Ann Arbor Science, Ann Arbor, MI.

Weeks, M. E. (1968). The discovery of the elements. Journal of Chemical Education, Inc. Easton, PA.

2

AQUEOUS GEOCHEMISTRY OF THALLIUM

Daniel I. Kaplan
Shas V. Mattigod

Pacific Northwest National Laboratory, P.O. Box 999, Mailstop K6-81, Richland, WA 99352

2.1. INTRODUCTION

Sources of thallium existing in the biosphere, that portion of the environment that impacts biota, can be broadly classified into two groups: natural sources and anthropogenic sources. Natural sources of thallium tend to be in a form that are less toxic than anthropogenic sources. Natural thallium sources do not exist in high concentrations (Shaw, 1952; Schoer, 1984; Kabata-Pendias and Pendias, 1984). Smith and Carson (1977) provided a thorough review

Thallium in the Environment, Edited by Jerome O. Nriagu.
ISBN 0-471-17755-5. © 1998 John Wiley & Sons, Inc.

of environmental concentrations of thallium and concluded that the crustal abundance of thallium ranges from 0.3 to 3 mg/kg, averaging about 1 mg/kg. A majority of the thallium in Earth's crust is carried in potassium minerals, such as alkali feldspars and micas or in association with sulfide minerals, such as pyrite (FeS_2) and sphalerite (ZnS) (Smith and Carson, 1977; Schoer, 1984) (Table 1). This can in part be attributed to the similar coordination number (>8) in silicates that Tl(I) and alkali metals have (Lee, 1971).

Only five independent thallium minerals have been reported: lorandite ($TlAsS_2$), vrbaite ($Hg_3Tl_4As_8Sb_2S_{20}$), hutchinsonite ($[Pb,Tl]_2[Cu,Ag]As_5S_{10}$), crookesite ($[Cu,Tl,Ag]_2Se$), and avicennite ($Tl_2O_3$). These minerals are very rare and are not the mining source for thallium. Thallium is generally mined from sulfide ores, such as galena (PbS), chalcopyrite ($CuFeS_2$), sphalerite (ZnS), and pyrite (Table 1). Deposits containing high concentrations of lead or arsenic are also often high in thallium (Shaw, 1952, 1957). Shales and manganese nodules may also contain high concentrations of thallium (Zitko, 1975).

The largest anthropogenic sources of thallium are coal combustion and heavy-metal (primarily zinc and cadmium) smelting and refining (Sager, 1994). Thallium, together with other volatile species tends to accumulate in filter and flue dust collected from combustion scrubbing processes. The thallium associated with the flue dust exists as sulfides that are remarkably soluble in water. Production and waste of electroplating and battery manufacturing are additional sources of thallium pollution. Thallium concentrations in the gram per liter range are often used as additives in electroplating of gold, chromium, nickel, lead, and zinc to improve adherence and uniformity of the coating. In battery electrolytes, thallium additives are used to prevent fowling of zinc anodes.

Table 1   Thallium Concentrations in Ores and Rocks

Mineral	Tl (mg/kg)	Reference
Galena (PbS)	1.4–20	Zitko, 1975
Sphalerite (ZnS)	8–45	Zitko, 1975
Pyrite (FeS$_2$)	5–23	Zitko, 1975
Calcareous-alkaline rocks	1.9–2.5	Shaw, 1952
		Smith and Carson, 1977
Granitic rocks	0.7–1.3	Shaw, 1952
		Smith and Carson, 1977
Sedimentary rocks	2.5	Shaw, 1952
		Smith and Carson, 1977
Marine sediments	0.2–5.7	Shaw, 1952
		Smith and Carson, 1977
Manganese nodules	1.9–199.8	Shaw, 1952
		Smith and Carson, 1977
Coal	0.05–10	Bowen, 1966
Limestone	0.5	Smith and Carson, 1977
Silica-carbonate soil	2.7	Smith and Carson, 1977

2.2. CHEMICAL AND PHYSICAL PROPERTIES OF THALLIUM

Thallium is the heaviest element of the IIIa subgroup of the periodic system; the IIIa group also contains boron, aluminum, gallium, and indium. Its periodic neighbors are mercury and lead. Thallium is not a transition metal. Elemental thallium is a soft metal. Its melting point and boiling point are relatively low (Table 2). Its low boiling point is an important property regarding its dispersion into the atmosphere. Because of its low boiling point, thallium is volatized during coal burning and several types of sheltering processes. Thus, if volatilized thallium escapes antipollution devices in smoke stacks, it can enter the environment in association with flue dust.

Thallium exists in two oxidation states: as $Tl(I)$ or thallous species or as $Tl(III)$ or thallic species. Thallium(I) compounds resemble quite closely the compounds of alkali metals and are soluble in water (Table 2). Thallium(I) compounds are easily oxidized by bromine, chlorine, hydrogen peroxide, or nitrous acid. Thallium(III) compounds are reduced to $Tl(I)$ compounds, for example, by sulfurous acid (Lee, 1971). Thallium(III) compounds tend to be appreciably more stable than $Tl(I)$ compounds.

Thallium(I) is commonly compared to potassium in the geochemistry literature; and for good reason, both ions are monovalent and they both have quite similar crystal radii [$Tl(I)$ = 1.40 Å; K^+ = 1.65 Å; Shannon, 1976]. Furthermore, $Tl(I)$ commonly replaces potassium in the crystal lattice of several minerals (Shaw, 1952; 1957; Smith and Carson, 1977). The reason these elements can interchange in the crystal lattice of minerals, that is, can undergo isomorphic substitution, is attributed to their similar ionic radii. However, ionic radius is only one of several parameters that must be taken into consideration when comparing the behavior of elements in the aqueous environment. Potassium is only a moderately good indicator of the complexation behavior of $Tl(I)$ in solution; rubidium and $Ag(I)$ are somewhat better indicators. However,

Table 2 Selected Physical and Chemical Properties of Thallium Important to Aqueous Geochemistry

Parameter	Units	Value
Density (20°C)	g/cm^3	11.85
Melting point	°C	302
Boiling point	°C	1453
Oxidation states	—	+1, +3
Tl^+ ionic radius	Å	1.40
Tl^{3+} ionic radius	Å	0.95
Tl^+ electron configuration	—	(Xe core)—$4f^{14}, 5d^{10}, 6s^2$
Tl^{3+} electron configuration	—	(Xe core)—$4f^{14}, 5d^{10}$

potassium will continue to be used by geochemists as an indicator of Tl(I) behavior because a great deal of information regarding potassium aqueous chemistry is available, and unlike rubidium and Ag(I), potassium is present in all natural systems. The main difference between the aqueous chemistry of Tl(I) and potassium can be attributed to their differences in electronegativity. Electronegativity is a measure of the power of an atom in a molecule to attract electrons to itself. Thallium(I) is appreciably more electronegative than any of the alkali elements. Associated with the increase of electronegativity from gallium to Tl(I), there is a marked increase in ionic radius of the M^+ ions. A similar trend is observed in Groups Ia (alkali elements), Ib, and IIb. Thus, one may expect to find greater similarities in the chemistries of cesium and Tl(I) than with potassium and Tl(I). Cesium is a "softer acid" than the other alkali elements (Huheey, 1983). Soft acids (or soft metals) tend to show a relatively greater affinity than hard acids for combining with more polarizable ligands compared with less polarizable ones. Hard acids have the inert-gas-type (d^0) electron configuration, that is, a spherical symmetry. Their electron sheaths are not readily deformed (nonpolarizable) under the influence of electric fields such as those produced by adjacent charged ions. They are, as it were, hard spheres, whereas soft acids have an electron sheath more readily deformable (polarizability) than that of hard acids and may be visualized as soft sphere (nd^{10} and nd^{10} $[n + 1]$ s^2 configurations). Polarizable species tend to form bonds that have a greater covalent character than nonpolarizable species. Thus, compared to potassium and the other alkali metals, Tl(I) bonds have a greater covalent character. The importance of the soft/hard (polarizable/nonpolarizable) designation is that soft acids tend to complex with soft bases whereas hard acids tend to complex with hard bases. Generally speaking, phosphorus is softer than nitrogen, sulfur is softer than oxygen, and iodine is softer than fluorine. Thus Tl(I) tends to form stronger complexes than potassium with phosphorus, sulfur, and iodine ligands. Conversely, potassium is more inclined to form complexes with nitrogen, oxygen, and fluorine ligands. The occurrence of thallium in a number of sulfide minerals (Table 1) is another indication of its soft acid nature. Importantly, these differences in the polarizability between thallium and potassium is appreciably more important in regards to aqueous complexation than to isomorphic substitution in minerals.

Thallium(III) is appreciably more electronegative than Tl(I). Thallium(III) is the most electronegative of the Group III elements. Associated with the increase of electronegativity from aluminum to thallium (as you move down the periodic chart), there is a marked increase in ionic radius of the M^{3+} ions. A similar trend is observed in Groups Ib and IIb, and one may thus expect to find similarities in the chemistries of Au(I), Hg(II), and Tl(III), all of which are soft acids (Lee, 1971).

2.3. OXIDATION–REDUCTION CHEMISTRY

The oxidation–reduction, or redox, limits of natural aqueous environments are defined by the redox couples of $O_2(g)/H_2O$ and $H_2(g)/H_2O$ (Table 3).

**Table 3   Electrochemical Properties
of Thallium**

Reaction	Standard Electrode Potential, V (25°C)
$\frac{1}{2}O_2 + 2e + 2H^+ = H_2O$	2.45
$H^+ + e = \frac{1}{2}H_2$	0.00
$Tl^+ + e = Tl$	-0.336
$Tl^{3+} + 2e = Tl^+$	1.25
$Tl(OH)_3 + 2e = TlOH + 2OH^-$	-0.05

When the redox potential is below the standard electrode potential of the $H_2(g)/H_2O$ couple, water is expected to be converted to H_2 gas. When the redox potential is above the standard electrode potential of the $O_2(g)/H_2O$ couple, water is expected to be converted to O_2 gas. Thus, natural aqueous systems are not expected to exist beyond these limits. The redox equations governing thallium are also presented in Table 3. As can be inferred from these standard electrode potentials, Tl(I) is likely to be the dominant species in aqueous environments. Only in the presence of extremely strong oxidizing agents (such as MnO_4^- and Cl_2) and high acidity would Tl(III) be expected to exist (Lee, 1971).

Observations of thallium behavior in the environment support these thermodynamic calculations (Shaw, 1952; Lanford, 1969; Wedepohl, 1972). The close correlation between thallium concentrations and rubidium and potassium concentrations in sediments and rocks lead Shaw (1952) to conclude that thallium existed primarily in these materials as a univalent species. Similarly, the adsorption of thallium to clays and its enrichment in sedimentary rocks formed in reducing conditions led Wedepohl (1972) to make the same conclusion. Both researchers speculated that thallium is probably transported in solution as Tl(I) and therefore is relatively quite mobile. Experiments conducted by Lanford (1969) further supported their conclusion. Lanford (1969) reported that treating a solution containing 2.4×10^{-4} M thallium with a stoichiometric amount of lime did not reduce the thallium concentration. This indicates that Tl(I) was the predominant species since Th_2O_3 would precipitate if it was Tl(III). When stoichiometric amounts of both lime and sulfide were added, the thallium concentration in solution decreased to 8.7×10^{-7} M, indicating 99.6% removal from solution.

2.4. AQUEOUS SPECIATION

This section will focus on the aqueous speciation of Tl(I), as compared to Tl(III), because it is the dominant species in the natural environment. Several of the stability constants for thallium complexes reported by Sillen and Martell

(1964), Lee (1971), and Baes and Mesmer (1976) are presented in Table 4. Few Tl(I) complexes are known and those that are known seldom have coordination numbers greater than 4. Thallium(I) complexes are generally neutral or anionic. Of the ligand important in water chemistry, thallium equilibrium constants for carbonate, phosphate, and fulvic acid have not been reported.

Complexes between most ligands and Tl(I) are relatively weak (Table 4). Weak complexation associated with Tl(I) can be attributed to the two alpha-antibonding electrons in the outer s orbital (Siegel, 1968). The hydroxide and chloride complexes of Tl(I) are all mononuclear (Baes and Mesmer, 1976). Though thallium complexes are rather weak, potassium can be displaced by thallium from essentially all organic ligand (Sager, 1994). This is of interest from the point of view of thallium toxicity to fish since humic acid complexes have been shown to decrease the toxicity of several metals, including Cu (Zitko, 1975).

Table 4   Stability Constants for Thallium Aqueous Complexes at 25°C

Complex	log K
$TlSO_4^-$	1–2
$TlS_2O_3^-$	1.91
$TlPO_4^{2-}$	3.14
$TlHPO_4^-$	1.2
$TlP_2O_4^{3-}$	4.01
$TlHP_2O_7^{2-}$	3.07
$TlNO_2^0$	0.80–0.85
$TlNO_3^0$	0.31–0.44
TlN_3^0	0.39
$TlNH_3^+$	-0.9 (2 mol/L NH_4NO_3)
$TlCl^0$	0.47–0.78
TlF^0	0.1
$TlOH^0$	0.42–0.85
$TlHC_2O_4^0$	2.03
$Tl(H_5C_6O_7)^{2-}$	0.65–1.04
$TlSO_4^+$	≈ 1
$Tl(SO_3)_4^{5-}$	≈ 34
$Tl(S_2O_3)_4^{5-}$	41
$TlCl^{2+}$, $Tl(Cl)_2^+$, $Tl(Cl)_3^0$, $Tl(Cl)_4^-$	8.1, 13.6, 15.8, 18.0
$TlOH^{2+}$	-0.2
$TlNO_2^{2+}$	0.18 (18°C)
$Tl(NH_3)_4^{3+}$	17
$Tl(HC_2O_4)_4^-$	≈ 15.4

Source: Sillen and Martell (1964), Lee (1971), and Baes and Mesmer (1976).

Thallium(III) forms much stronger complexes than Tl(I). The Tl(III) inorganic complexes are generally of the form of TlX_4^- or TlX_6^{3-} (X = halogen, sulfate, nitrate, acetate, etc.; Table 4). Wade and Banister (1973) concluded from their review of 4-, 5-, and 6-coordinate Tl(III) halo complexes that among the most stable metal chloride complexes are $TlCl^{2+}$, $TlCl_2^+$, and $TlCl_4^-$. The hydroxide and chloride complexes of Tl^{3+} are all mononuclear (Baes and Mesmer, 1976). The hydrolysis of Tl^{3+} begins at very low pH values. Thallium(III) hydroxide is rather insoluble in water.

The Tl(III) organic complexes are generally of the form $RTlX_2$, R_2TlX, and R_3Tl (R = organic group; Smith and Carson, 1977; O'Shea and Mancy, 1978). Of the trivalent organothallium compounds, the R_2TlX compounds are the most stable. In aqueous solutions, most of them dissociate into R_2Tl^+ and X^-. Water solubility is a function of R and X and very likely decreases with increasing hydrophobicity of R.

Some understanding of aqueous speciation of Tl(I) in natural systems can be gained by conducting illustrative computations using one of the available speciation models (GEOCHEM-PC, Parker et al., 1995). We chose the estimated compositions of several natural water systems representing a range of conditions of pH, Tl(I) concentrations, dissolved organic carbon, and major metal and ligand concentrations. Typical total concentrations of Tl(I) and dissolved organic carbon typically encountered in these water systems were obtained from the compilations of Smith and Carson (1977) and Thurman (1985), respectively. The estimated mean compositions of aqueous systems used in our computations are listed in Table 5.

Most association constants used in our calculations for Tl(I) complexes were obtained from Sillen and Martell (1972) (Table 6). Thallium complexation constants for carbonate, phosphate, and fulvic acid are not reported in the literature. Thus, these constants were calculated. Values for $TlHCO_3^0$ and $TlCO_3^-$ were estimated from the method of Langmuir (1979). The association constant for $TlH_2PO_4^0$ was estimated as an average of log K ratios of PO_4^{3-}/$H_2PO_4^-$ and HPO_4^{2-}/$H_2PO_4^-$ for various metal phosphate complexes. Association constants for Tl(I) complexes with dissolved organic matter (as fulvate complexes) were calculated according to the method of Sposito (1981). Sposito (1981) described metal complexation to fulvic acid as occurring at two sites, FUL1 and FUL2. The FUL1 site approximates the carboxyl functional groups on fulvic acids, whereas the FUL2 sites approximate the hydroxyl functional groups on fulvic acids.

The computations were conducted using the speciation model GEOCHEM-PC (Parker et al., 1995). The computations for these 6 metal and 10 ligand systems involved 64 aqueous complex species. The results of the computation showed that a major fraction (about 90%) of the total dissolved Tl(I) in a typical groundwater would exist in free ionic (uncomplexed) form with minor fractions ($<5\%$ each) as complexes with bicarbonate and sulfate ligands (Table 7). In typical river water, the free ionic form would constitute about 83% of the total dissolved thallium, whereas about 16% would exist as an organic

Table 5 Estimated Mean Composition (mg/L) of Some Natural Water Systems[a]

Dissolved Constituent	Groundwater	River Water	Eutrophic Lake Water	Bog Water	Seawater
Ca	59	15	40	0.2	422
Mg	26	4.1	10	0.19	1322
Na	22	6.3	9	1.5	11020
K	4	2.3	2	0.31	408
Tl(I)	0.00725	0.00002	0.00002	0.00002	0.000013
CO_3	266	57	122	0.06	145
SO_4	108	11	24	0.53	2775
Cl	11	7.8	9	0.99	19805
F	0.1	1	1	0.1	1.4
NO_3	39	1	3.4	1	0.3
PO_4	0.1	0.0767	0.6	0.0767	0.0614
H_4SiO_4	48	20.8	2	20.8	4.4
Organic Carbon	0.7	5.0	10.0	30.0	0.5
pH	7.14	8.01	7.70	3.60	8.22

[a] References for water composition: Groundwater and lake water (Stumm and Morgan 1981); river water (Hem 1985); bog water (Thurman 1985); seawater (Nordstrom et al. 1979); Tl(I) (Smith and Carson 1977); organic carbon (Thurman 1985).

22

Table 6   Association Constants (log K) for Tl(I) Complex Species Used in Speciation Calculations

Species	log K
TlOH^0	0.79
TlHCO$_3$^{0}	1.2
TlCO$_3$$^-$	2.25
TlSO$_4$$^-$	1.8
TlCl^0	0.49
Tl(Cl)$_2$$^-$	0.0
TlF^0	0.1
TlNO$_3$^{0}	0.33
TlH$_2$PO$_4$^{0}	0.68
TlHPO$_4$$^-$	1.2
TlPO$_4$$^{2-}$	3.14
Tl-Ful1^0	4.83
Tl-Ful2^0	3.32

complex (Tl–fulvate). Computations showed that in typical eutrophic lake waters, almost a fifth of dissolved thallium would exist as organic complexes. In typical bog waters, the organic bound fraction constitutes a significant fraction (about 67%) of total dissolved thallium. In some bogs, with higher concentrations of dissolved organics (400 mg/L), almost all thallium ($\sim$96%) would exist as organically complexed species.

Speciation calculations indicated that in typical seawater, only about 52% of total dissolved thallium would exist in free ionic form whereas about 36 and 11% would be complexed with chloride ($\sim$31% as TlCl^0 and $\sim$5% as Tl[Cl]$_2$$^-$) and sulfate ligands, respectively. These computations indicated that

Table 7   Computed Distribution of Tl(I) Species (% of total) in Selected Natural Water Systems^a

Aqueous Species	Groundwater	River Water	Eutrophic Lake Water	Bog Water	Seawater
Tl$^+$	90.4	82.7	76.8	32.4	51.9
TlHCO$_3$^{0}	4.4	1.2	2.0	—	0.5
TlCO$_3$$^-$	—	—	—	—	0.1
TlSO$_4$$^-$	3.6	0.4	0.8	—	11.2
TlCl^0	0.1	0.1	0.1	—	30.7
Tl(Cl)$_2$$^-$	—	—	—	—	5.4
Tl-Fulvate^0	1.4	15.6	20.3	67.6	0.2

^{a} Concentrations of TlOH^0, TlF^0, TlNO$_3$^{0}, TlH$_2$PO$_4$^{0}, TlHPO$_4$$^-$, and TlPO$_4$$^{2-}$ species were negligible in all systems.

under typical pH conditions encountered in surface waters (pH 7–8), aqueous Tl(I) would exist primarily as free ionic species (77–90%). However, in highly acidic bog waters, organically complexed thallium would constitute the major fraction (68–96%) of dissolved Tl(I). In seawater inorganic complexes of Tl(I) was predicted to constitute almost one half of dissolved Tl(I), indicating that in other inorganic-ligand-dominated systems, such as alkaline lake and brines, inorganic-ligand-bound Tl(I) would dominate Tl(I) dissolved speciation.

2.5. PRECIPITATION AND DISSOLUTION

The solubility of several Tl(I) compounds are presented in Table 8. These values, like those for alkali metals, are quite high, suggesting that Tl(I) would not precipitate from solution in most environments. However, in the trivalent state, quantitative coprecipitation with hydroxides of aluminum, iron, manganese, magnesium, and zirconium has been reported (Shaw, 1952). The coprecipitation of thallium with manganese hydroxide can lead to a significant enrichment of thallium in deep-sea ferromanganese nodules, corresponding to a strongly positive correlation with their manganese contents, but none with aluminum or potassium of geologic materials (Shaw 1952).

Table 8   Solubility of Selected Tl I) Compounds

Tl(I) Compound	Solubility (g/L)
Aluminum sulfate	117.8
Carbonate	40.3
Chloride	2.9
Cyanide	168
Ferrocyanide	3.7
Fluoride	786
Formate	5000
Hydroxide	259
Iodide	0.006
Nitrate	95.5
Nitrite	321
Oxalate	14.8
Orthophosphate	5
Pyrophosphate	400
Sulfate	48.7
Sulfide	0.2
Tetraphenylborate	0.0005
Dithionate	418
Salts of fatty acids	0.1–1

Source: Zitko (1975) and Sager (1994).

2.6.   ADSORPTION AND DESORPTION

In studying thallium adsorption to geologic materials, it is very important to differentiate between thallium derived from anthropogenic and natural sources. The thallium from anthropogenic sources tends to be in an appreciably more easily extractable fraction than natural thallium (Lehn and Schoer, 1987; Sager, 1992). The cause for this is that the thallium of natural origin tends to be incorporated into the structure of the solid phase, whereas thallium of anthropogenic sources tends to be adsorbed onto the surface of the minerals on easily exchangeable sites. For example, in uncontaminated sediments, Lehn and Schoer (1987) found that a significant portion of the thallium was in the organic or the residual (crystalline structural) fractions. However, in thallium-contaminated sites, the thallium was almost entirely in the exchangeable fraction (Lehn and Schoer, 1987; Sager, 1992). Sager (1992) reported that >95% of the adsorbed thallium on two river sediments was in an easily exchangeable form (exchangeable with 1 M ammonium acetate, pH 7). Lehn and Schoer (1987), using a similar extractant, reported similar results with soils containing vermiculite minerals.

The studies of Lehn and Schoer (1987) and Sager (1992) showed that essentially all of the recently adsorbed thallium could be desorbed. This is consistent with a cation exchange adsorption mechanism, as opposed to a precipitation, coprecipitation, or absorption (into the structure of the solid phase) mechanism. Cation exchange, as the name implies, is the reversible adsorption reaction in which an aqueous species exchanges with an adsorbed species. Cation exchange reactions are approximately stoichiometric and can be written, for example, as

$$KX(s) + Tl^+(aq) = TlX(s) + K^+(aq) \qquad (1)$$

where X designates and exchange surface site. The adsorption of potassium and rubidium is also controlled by cation exchange reactions.

Thallium(I) adsorption onto micaceous minerals (layered silicates with a high potassium adsorption potential) is unique in that it does not follow the cation exchange reaction (Frantz and Carlson, 1987). Instead Tl(I) adsorbs onto micaceous minerals by the cation fixation mechanism. This type of adsorption occurs when a cation is just the right size to fit between the individual layers making up the mica mineral. This type of adsorption is very strong and essentially irreversible. Kittrick (1966, 1969) suggested the occurrence of cation fixation is primarily a function of the extent to which a cation can shed its hydration shell before entering into a mineral interlayer. Those cations with hydration energies less negative than that of potassium will shed their hydration shells upon entering interlayer spaces and become fixed (strongly adsorbed), while those cations with hydration energies more negative than potassium will remain hydrated and thus impeded from entering the interlayer spaces. Rubidium, cesium, and thallium have hydration energies less negative than potassium (Hunt, 1963). The abundance of these elements in micaceous minerals decreases in the order

$$Rb > Cs > Tl. \tag{2}$$

The relative tendency of these element to be adsorbed into clays has been give by Canney (1952) to be

$$Cs > Rb > Tl. \tag{3}$$

The differences between Eqs. (2) and (3) can be attributed to the differences in crustal abundance of these elements.

There has not been many field experiments evaluating the propensity of thallium to adsorption to soils. Heinrichs and Mayer (1977) evaluated the annual fluxes of thallium precipitation in a Central European beach and spruce forest. Thallium generally passed through the forest canopy. Once through the canopy, it accumulated in the organic surface horizon and did not move through to the lower soil horizons. This suggested to the authors that thallium has a low mobility in the high cation exchange soils at the study site. In a study in which thallium salts were added to a highly weathered soil in South Carolina, thallium was also reported not to move readily through the soil column during a 30-month period (Martin and Kaplan, 1997; Kaplan et al., 1990). Furthermore, they reported that bioavailability of thallium decreased over time as a result of soil transformation into unavailable chemical forms and not to leaching. These changes in bioavailability occurred soon after application of the thallium salt, becoming negligible after 18 months. Iron and aluminum oxides were responsible for retaining the thallium in the upper portion of the soil profile.

2.7. CONCLUSIONS

Thallium enters the biosphere from natural and anthropogenic sources. Natural sources of thallium are less bioavailable and therefore of less concern to regulators than anthropogenic sources of thallium. Thallium crustal abundance ranges from 0.3 to 3 mg/kg, averaging about 1 mg/kg (Smith and Carson, 1977). The largest anthropogenic source of thallium is related to coal combustion and heavy-metal (primarily zinc and cadmium) smelting and refining (Sager, 1994).

Thallium can exist in nature in two oxidation states, $+1$ and $+3$. Thallium is more commonly found in the lower oxidation state in aqueous environments. Its geochemical behavior is quite similar to that of potassium and rubidium, two alkali metals that also have an oxidation state of $+1$. Thallium(I) has a similar ionic radius to potassium and rubidium, permitting it to substitute for these elements into the crystal structure of several minerals (Shaw, 1952; Smith and Carson, 1977). However, the aqueous chemistry of Tl(I) does not completely parallel that of rubidium and potassium due to its softer, more polarizable, bonding properties. An example of the importance of this difference is that thallium, compared to potassium or rubidium, forms stronger sulfide and aromatic organic compound complexes (Sillen and Martell, 1964).

Few Tl(I) complexes are known; those that are know, are relatively weak. Furthermore, several Tl(I) complexation constants have not been measured, including those for carbonate, phosphate, and fulvic acids. Thus, a complete Tl(I) speciation in natural systems has never been conducted. For this chapter, these constants were estimated using techniques reported in the literature. The calculated association constants (log K) for $TlH_2PO_4^0$ was 0.68, for $TlHPO_4^-$ was 1.2, for $TlPO_4^{2-}$ was 3.14, for $TlHCO_3^0$ was 1.2, for $TlCO_3^-$ was 2.25, for fulvic acid(I)^0 was 4.83, and for fulvic acid(II)^0 was 3.32. The addition of these species into the speciation calculation for a number of "typical waters" provided a more complete representation of the speciation. Furthermore, some important observations were made, especially in sea water and bog water, that have in the past been overlooked by researchers doing similar calculations without these species included in their calculations.

Speciation computations predict that the most common form of Tl(I) in ground, river, and lake waters would be the free Tl^+ species. However, the computations also predicted that in other natural water systems such as bogs, organically complexed Tl(I) would be the dominant form. This is of interest from the point of view of thallium toxicity to fish since organic complexes of metals have been shown to decrease fish toxicity (Zitko, 1975). In high ionic strength water systems, such as seawater, alkaline lake water, and brines, an significant fraction of Tl(I) would be bound up as inorganic complexes, primarily as $TlSO_4^-$ and $TlCl^0$. Thus, in high organic matter or high ionic strength water, Tl(I) would be expected to exist as complexes.

The solubility of Tl(I) compounds, like those for alkali metals, are quite high. This suggests that Tl(I) would not precipitate from solution in most environments. However, in the rare instance when Tl(III) exists, thallium may precipitate from solution.

In understanding thallium adsorption to geologic materials, it is very important to differentiate between thallium derived from anthropogenic and natural sources. The thallium from anthropogenic sources tends to be in an appreciably more easily extractable fraction than natural thallium. The reason for this is that the thallium of natural origin tends to be incorporated into the structure of the solid phase, whereas thallium of anthropogenic sources tend to be adsorbed onto the surface of the minerals on easily exchangeable sites. The dominant mechanism of thallium adsorption appears to be cation exchange, as opposed to precipitation, coprecipitation, or absorption into the structure of the solid phase. Cation exchange is a process that does not bond the Tl(I) very tenaciously. Thus, thallium would be expected to move rather readily with groundwater flow, that is, its attenuation by sediments would be rather limited.

REFERENCES

Baes, C. R., and Mesmer, R. E. (1976). *The Hydrolysis of Cations*. Wiley, New York, pp. 328–335.

Bowen, H. J. M. (1966). *Trace Elements in Biochemistry*. Academic, New York, p. 216.

Canney, F. C. (1952). Some aspects of the geochemistry of K, Rb, Cs, and Tl in sediments. *Geol. Soc. Am. Bull.*, **63**, 1238–1239.

Frantz, G., and Carlson, R. M. (1987). Effects of rubidium, cesium, and thallium on interlayer potassium release from Transvaal vermiculite. *Soil Sci. Soc. Am. J.* 51, 305–308.

Heinrichs, J., and Mayers, R. (1977). Distribution and cycling of major and trace elements in two Central European forest ecosystems. *J. Environ. Qual.*, **6**, 402–407.

Hem, J. D. (1985). *Study and Interpretation of the Chemical Characteristics of Natural Water*. U.S.G.S. Water Supply Paper 2254, U.S. Geological Survey, Alexandria, Virginia.

Huheey, J. E. (1983). *Inorganic Chemistry, 3rd ed.* Harper & Row, New York.

Hunt, J. P. (1963). *Metal Ions in Aqueous Solution*. W. A. Benjamin, New York, p. 321.

Kabata-Pendias, A., and Pendias, H. (1984). *Trace Elements in Soils and Plants*. CRC Press, Boca Raton, FL, pp. 138–139.

Kaplan, D. I., Adriano, D. C., and Sajwan, K. S. (1990). Thallium toxicity in bean plants. *J. Environ. Qual.*, **19**, 359–365.

Kittrick, J. A. (1966). Forces involved in ion fixation by vermiculite. *Soil Sci. Soc Am Proc.*, **30**, 801–803.

Kittrick, J. A. (1969). Interlayer force in montmorillonite and vermiculite. *Soil Sci. Soc. Am. Proc.*, **33**, 217–225.

Lanford, C. (1969). Effect of trace metals on stream ecology. Presented at the 1969 Cooling Tower Institute Annual Meeting, Anaheim, California, January 20, 1969.

Langmuir, D. (1979). Techniques of estimating thermodynamic properties for some aqueous complexes of geochemical interest. In *Chemical Modeling in Aqueous Systems*, Jenne, E. A. (ed.). American Chemical Society Symp. Series 93, American Chemical Society, Washington, DC, pp. 353–387.

Lee, A. G. (1971). *The Chemistry of Thallium*. Elsevier, Amsterdam, p. 336.

Lehn, H., and Schoer, J. (1987). Thallium-transfer from soils to plants: Correlation between chemical form and plant uptake. *Plant Soil*, **97**, 253–265.

Martin, H. W., and Kaplan, D. I. (1997). Temporal changes in cadmium, thallium, and vanadium mobility in soil and bioavailability under field conditions. *Water Air Soil Poll.* (in press).

Nordstrom, D. K., Plummer, L. N., Wigley, T. M. L., Wolery, T. J., Bau, J. W., Jenne, E. A., Basset, R. L., Crerar, D. A., Florence, T. M., Fritz, B., Hoffman, M., Holdren, G. R., Lafon, G. M., Mattigod, S. V., McDuff, R. E., Morel, F., Reddy, M. M., Sposito, G., and Thraikill, J. (1979). Comparison of computerized chemical models for equilibrium calculations in aqueous systems. In *Chemical Modelling in Aqueous Systems*, Jenne, E. A. American Chemical Society Symp. Series 93, American Chemical Society, Washington, DC, pp. 857–894.

O'Shea, T. A., and Mancy, K. H. (1978). The effect of pH and hardness metal ions on the competitive interaction between trace metal ions and inorganic and organic complexing agents found in natural waters. *Water Res.*, **12**, 703–711.

Parker, D. R., Norvell, W. A., and Chaney, R. L. (1995). GEOCHEM-PC, A chemical speciation program for IBM and compatible personal computers. In *Chemical Equilibrium and Reaction Models*, Loeppert, R. H., Schwab, A. P., and Goldberg, S. (eds). Soil Science Society of America, Special Publication 42, Soil Science Society of America, Madison, WI, pp. 245–277.

Sager, M. (1992). Speciation of thallium in river sediments by consecutive leaching techniques. *Mikrochim. Acta*, **106**, 241–251.

Sager, M. (1994). Thallium. *Toxicol. Environ. Chem.*, **45**(1/2), 11–32.

Schoer, J. (1984). Thallium. In *Handbook of Environmental Chemistry*, Vol. 3c, Hutzinger, O., (ed.). Springer, Berlin, pp. 143–214.

Shannon, R. D. (1976). Revised effective ionic radii and systematic studies of interatomic distances in halides and chalcogenides. *Acta Cryst.*, **A32**, 751–767.

Shaw, D. M. (1952). The geochemistry of thallium. *Geochim. Cosmochim. Acta*, **2**, 118–154.

Shaw, D. M. (1957). The geochemistry of gallium, indium, and thallium—A review. *Phys. Chem. Earth*, **2**, 164–211.

Siegel, B. (1968). Oxyhalides of boron, aluminum, gallium, indium, and thallium. *Inorg. Chim. Acta Rev.*, **2**, 137–146.

Sillen, L. G., and Martell, A. E. (1964). *Stability Constants of Metal-Ion Complexes*, Special Publication Number 17, The Chemical Society, London, p. 754.

Smith, I. C., and Carson, B. L. (1977). *Trace Metals in the Environment*, Vol. 1, *Thallium*, Ann Arbor Scientific, Ann Arbor, MI, pp. 138–139.

Sposito, G. (1981). Trace metals in contaminated waters. *Environ. Sci. Tech.*, **15**, 396–403.

Thurman, E. M. (1985). *Organic Geochemistry of Natural Waters*. Martinus Nijhoff/Dr W. Junk Publishers, Dordrecht, Netherlands.

Wade, K., and Banister, A. J. (1973). *The Chemistry of Aluminum, Gallium, Indium, and Thallium*, Vol. 12. Pergamon Texts in Inorganic Chemistry, Pergamon Press, New York, p. 214.

Zitko, V. (1975). *Chemistry, Application, Toxicity and Pollution Potential of Thallium*. Technical Report Number 518, Department of the Environment, Fisheries and Marine Service, Research and Development Directorate, Biological Station, St. Andrews, New Brunswick, Canada, p. 41.

3

SPECIATION OF THALLIUM IN NATURAL WATERS

Tser-Sheng Lin and Jerome O. Nriagu

Department of Environmental and Industrial Health, School of Public Health, University of Michigan, Ann Arbor, MI 48109

3.1. INTRODUCTION

Little is currently known about the behavior and fate of thallium in lacustrine environments (also see Chapters 2 and 4). In fact, few people could detect thallium in water samples until fairly recently (Flegal and Patterson, 1985). Measurements obtained during the past few years show that dissolved thallium concentrations average 10–15 ng/kg in seawater and 5–10 ng/L in unpolluted and 20–50 ng/L in polluted freshwater samples (Flegal and Patterson, 1985; Cheam et al., 1995). The low concentrations especially in unpolluted environments entail severe methodological difficulties in trying to determine the forms

Thallium in the Environment, Edited by Jerome O. Nriagu.
ISBN 0-471-17755-5. © 1998 John Wiley & Sons, Inc.

of dissolved thallium in a given water sample. In the absence of any actual measurements, a vacuum currently exists on what we know about the speciation of thallium in natural waters. Thermodynamically derived distributions of aqueous thallium species are often contradictory because of the large uncertainty in the reported association constants. In this chapter, we use the most recent and what we consider to be a consistent set of thermodynamic data to model the speciation of thallium in natural waters. The stability relationships should provide a framework and serve as a guide in future attempts to determine the forms of this element in natural waters. This chapter fills the gap between the discussions of the stability of minerals and solid phases (Chapter 4) and the distribution and fate of thallium in aqueous environments (Chapter 2).

3.2. HYDROLYSIS

The hydrolysis of thallium is controlled by the pH of the aqueous solutions. Reported hydrolysis constants (K_h) for Tl(III) are as follows:

$$Tl^{3+} + OH^- = Tl(OH)^{2+}; \quad K_1 = 10^{11.31}$$

$$Tl(OH)^{2+} + OH^- = Tl(OH)_2^+; \quad K_2 = 10^{7.64}$$

$$Tl(OH)_2^+ + OH^- = Tl(OH)_3^0; \quad K_3 = 10^{6.58}$$

$$Tl(OH)_3^0 + OH^- = Tl(OH)_4^-; \quad K_4 = 10^{5.22}$$

The calculated distribution of these hydrolysis products is shown in Figure 1 as a function of pH. The predominant Tl(III) hydroxyl species is $Tl(OH)^{2+}$

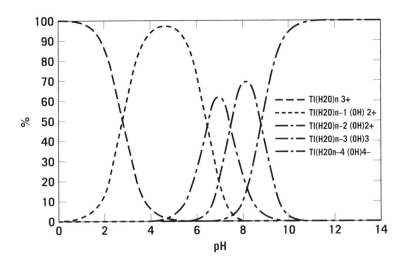

Figure 1. Distribution of Tl(III) hydroxyl species as a function of pH.

Table 1 Solubility of Thallium Compounds

Compound	Solubility g/L (20°C)	Solubility Product (log K_{sp})
Thallium bromide	0.48	−5.59
Thallium carbonate	52.3	−3.82
Thallium(I) chloride	3.4	−13.42
Thallium(III) chloride	826	
Thallium ferrocynaide	3.7	−7.44
Thallium fluoride	786	
Thallium(I) hydroxide	350	
Thallium iodine	0.06	
Thallium nitrate	86.7	
Thallium oxalate	15.3	
Thallium phosphate	5	−8.18
Thallium sulfate	46.4	
Thallium sulfide	0.2	
Thallium(III) hydroxide		−45.2

when pH ranges from 4 to 6, $Tl(OH)_2^+$ becomes the dominant form when pH is around 7, and $Tl(OH)_3$ will dominate when pH is in the range of 7.5–8.8. In contrast to the distribution of Tl(III) hydroxyl complexes, Tl^+ ($K_h = 10^{-11.7}$) is the predominant form of Tl(I) expected in the pH range likely to be encountered in most natural waters.

Although most of the trivalent thallium compounds are soluble, the hydroxide [$Tl(OH)_3$] (log $K_{sp} \sim -45.2$) and oxide are insoluble (Table 1). Compared to other trivalent metal hydroxides, $Tl(OH)_3$ has the lowest K_{sp} value (Table 2). Thus, under oxidizing conditions, the formation of solid $Tl(OH)_3$ may be expected to control the level of Tl in the aqueous environment. Calculation of solubility of Tl(III) hydroxide as a function of pH shows that in the absence of other complexing ligands, the total concentration of dissolved Tl(III) is predicted to be very low when the pH > 3.0. By contrast, the solubility of monovalent thallium hydroxide, $Tl(OH)_{(solid)}$ is reported to be 350 g/L (Dean,

Table 2 Solubility Products of Trivalent Metal Hydroxides

Species	pK_{sp}
$Cr(OH)_3$	30.2
$Fe(OH)_3$	37.4
$Al(OH)_3$	32.9
$Ga(OH)_3$	35.2
$In(OH)_3$	33.2
$Tl(OH)_3$	45.2

Source: Dean, 1985.

1985) and thus is unlikely to be precipitated under normal environmental conditions.

3.3. COMPLEXATION

The Tl^+ ion is similar to both the alkali metals and the Ag^+ ion in its chemistry (Baes and Mesmer, 1976; Lee, 1971). Like the alkali metal ions, Tl^+ forms relatively few strong complexes (Cotton and Wilkinson, 1972). On the other hand, monovalent thallium resembles silver in forming an insoluble sulfide (solubility product for Tl_2S is $10^{-21.2}$) and sparingly soluble halide complexes that are light sensitive (Baes and Mesmer, 1976; Lee, 1971).

As expected, Tl(III) forms much more stable complexes than Tl(I) does (Table 3). Figure 2 is a predominance diagram for Tl^{3+}–OH^-–Cl^- species at varying concentrations of $[Cl^-]$. This figure predicts that $Tl(OH)Cl_2$ is the predominant form of Tl(III) in natural waters. Furthermore, the solubility of $Tl(OH)_3$ at $[Cl^-] = 1 \times 10^{-4} M$ and pH $= 8$ is increased by about 1000-fold due to the formation of the $TlOHCl_2$ complex. The important thing to note is that Tl(III) can be stabilized by the presence of chloride ions (Cotton and Wilkinson, 1972) although Tl(III) is easily reduced to Tl(I) in aqueous solution ($Tl^{3+} + 2 e^- \leftrightarrow Tl^+$, log $K = 43.3$).

Using the MINEQL equilibrium model (version 3), the distributions of thallium species in the presence of sulfate and bromide ions have been calculated (Figs. 3 and 4). The calculations show that thallium–bromide or thallium–sulfate complexes are more stable than (should predominate over) Tl^{3+} ions in acidic aqueous environments. In an open system with carbonate ions, the MINEQL model also predicts that $TlCO_3^+$ would be the predominant form of Tl(III) species in the pH range of 0–14. The interaction of thallium and iodine is interesting. Since the iodide–iodine standard potential (-0.53 V) is much higher than that of the Tl(I)–Tl(III) couple (-1.28 V), the reaction $Tl^{3+} + 2I^- \leftrightarrow Tl^+ + I_2$ can occur spontaneously. The equilibrium constant is about 2×10^{25}. The reaction will likely shift to $Tl^{3+} + 4I^- \leftrightarrow TlI_4^-$ in which the association constant is about 4.6×10^{35}. The ratio of iodide–thallium may thus be a critical factor in the distribution of Tl in some natural waters.

3.4. SURFACE AND ADSORBED SPECIES

In addition to dissolved species and solid phases, the adsorbed forms represent another important Tl component in natural waters. Various models have been used to describe the adsorption/desorption of trace metals on suspended particulate matter in natural waters. In general, most of the experimental data can be fit to a generalized two-layer model (Dzombak and Morel 1990). The two-layer model treats the sorption of solutes at oxide surfaces as a chemical reaction; with specific surface sites the sorption constant for metal at oxide

Table 3   Stability Constants for Thallium Complexes

Ligand	Number of Ligand	Tl(I) log K	Tl(III) log K
Citric acid	1	1.04	12.02
NTA	1	3.44	16.81
Acetate	1	0.79	8.42
Br^-	1	0.93	9.70
Br^-	2		16.60
Br^-	3		21.20
Br^-	4		23.90
Cl^-	1	0.52	8.14
Cl^-	2		13.60
Cl^-	3		15.78
Cl^-	4		18.00
EDTA	1		22.50
F^-	1	0.10	6.44
I^-	1	0.72	11.42
I^-	2		20.88
I^-	3		27.60
I^-	4		31.82
Nitrate	1	0.33	0.92
$SO_4^=$	1	0.95	9.02
$HPO_4^=$	1	3.31	17.66
$SO_4^=$	2	1.02	9.28
$NO3-$	1	0.45	7.20
$CO_3^=$	2	2.79	15.76
$HCO3-$	1	3.42	18.07
Sorption-natural particle	1	2.50	14.68
$\equiv S\text{-}OH_{weak}$	1	0.24	6.41
$\equiv S\text{-}OH_{strong}$	1	1.31	10.35
Biosorption-S-site	1	1.25	10.83
Biosorption-C-site	1	1.62	12.10

surfaces can be estimated by linear-free-energy relationships if it has not been determined empirically. The adsorption constants for Tl(I) and Tl(III) have been so estimated and are shown in Table 3. The effects of adsorption of Tl(I) and Tl(III) on ferric oxide surfaces on thallium speciation in natural waters have been models using MINEQL (Figs. 5 and 6). The adsorption reaction, $\equiv S\text{—}OH$ (ferric oxide surface) $+Tl^+ \rightarrow\ \equiv S\text{—}OTl + H^+$, starts to exert a significant influence on Tl^+ concentration at pH of 3, and the Tl^+ will be completely adsorbed at pH of about 6.5. On the other hand, the adsorption of Tl(III) according to the scheme $\equiv S\text{—}OH + Tl^{3+} \rightarrow\ \equiv S\text{—}OTl^{2+} + H^+$ becomes significant at pH of 4.6, and at pH of about 6.5 most of the dissolved Tl^{3+} would be completely adsorbed. Figure 7 compares the adsorption edges

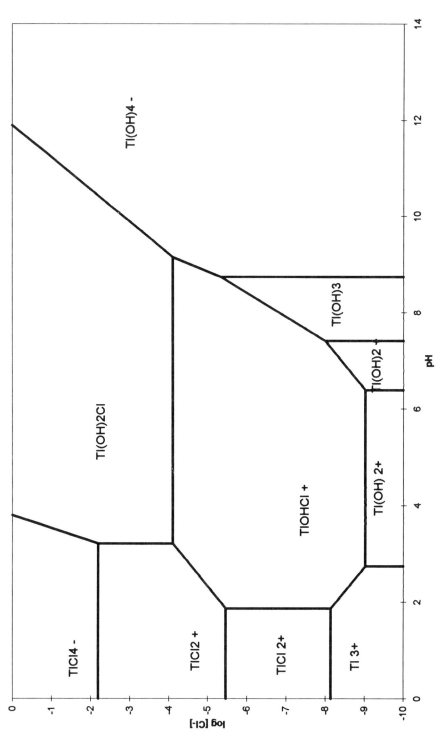

Figure 2. Calculated predominance diagram for dissolved species in the Tl^{3+}–OH^-–Cl^- system.

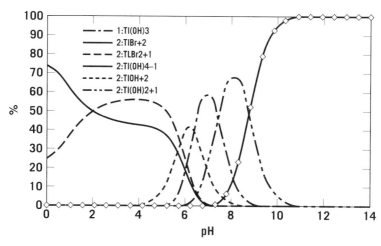

Figure 3. Distribution of monomeric hydroxyl and bromide complexes of thallium as a function of pH. Total bromide concentration $= 1.75 \times 10^{-7}$ M; total Tl^{3+} concentration $= 1 \times 10^{-10}$ M; precipitation of $Tl(OH)_3$ is neglected.

of Cu^{2+}, Pb^{2+}, Cd^{2+}, Zn^{2+}, Tl^{3+}, and Tl^+ derived using the two-layer model for ferric oxide surfaces at various pH. The important thing to note is that adsorption of Tl^+ occurs at lower pH range compared to other metals, because Tl^+ does not undergo significant hydrolysis in the acidic solutions.

Passive nonmetabolic process of metal ion binding by living or dead biomass, or biosorption, has been described successfully by MINEQL program for *Sargassum fluitans* (Schiewer and Volesky, 1996). There are two binding

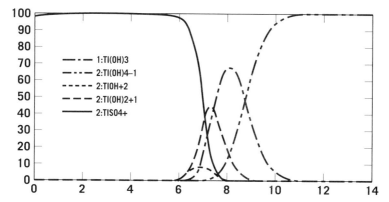

Figure 4. Distribution of hydroxyl and sulfato complexes of thallium as a function of pH. Total sulfate concentration $= 1 \times 10^{-4}$ M; total Tl^{3+} concentration $= 1 \times 10^{-10}$ M; precipitation of $Tl(OH)_3$ is ignored.

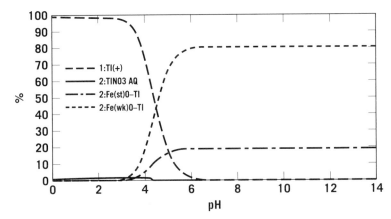

Figure 5. Predicted adsorption of Tl(I) by hydrous ferric oxide using the two-layer model.

sites in *Sargassum* biomass—the carboxyl groups of alginate and the sulfate groups of fucoidan (Schiewer and Volesky, 1996). Estimated adsorption constants of *S. fluitans* for Tl(I) and Tl(III) (Table 3) are greater than those of hydrous ferric oxide. Biosorption conceivably should be an important phenomenon that influences the speciation for thallium in natural waters. The study by Muller and Sigg (1990) showed that the adsorption of trace metals on natural particle surfaces was much stronger than those on hydrous ferric oxide surfaces. For instance, the constant for the adsorption of lead on natural particle surfaces was found to be $10^{9.44}$ compared to $10^{4.65}$ for adsorption on hydrous ferric oxide surfaces. The higher adsorption capacity is not surprising;

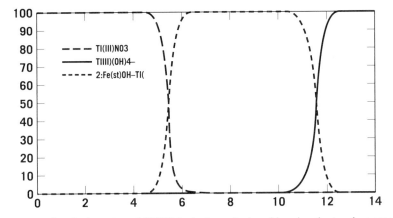

Figure 6. Predicted adsorption of Tl(III) by hydrous ferric oxide using the two-layer model.

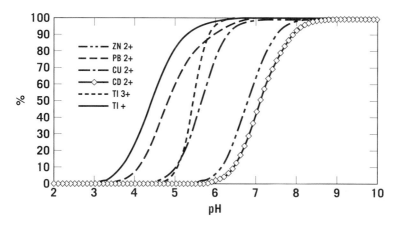

Figure 7. Comparison of the adsorption of Pb, Cu, Zn, Cd, Tl(I), and Tl(III) by a hydrous ferric oxide using the two-layer model.

suspended particulates in water often consist of a mixture of inorganic and organic chelating ligands.

Thallium(I) readily replaces potassium in clays and secondary silicate minerals because of its greater electronegativity compared to K(I) (Magorrian et al., 1974). On the other hand, Tl(III) is primarily associated with iron/manganese oxyhydroxides. The reported concentration of Tl in marine manganese nidules range from 2300 to 91000 μg/g (Cronan, 1976). The atomic ratio of thallium to potassium in ferromanganese nodules (4×10^{-3}) is 6 orders of magnitude greater that that in seawater (5×10^{-9}) (Cronan, 1976). Moreover, the thallium concentration in manganese-rich sediment (1.67 μg/g) is 5 orders of magnitude greater than that in seawater (18.7 ng/L) (Riley and Siddiqui, 1986). A study by Bidoglio et al. (1993) suggests that thallium is accumulated by mineral oxides as a result of Tl(I) sorption followed by oxidation at the mineral surfaces.

3.5. OXIDATION AND REDUCTION

Oxidation–reduction reactions exercise a controlling influence on chemical speciation thallium that can exist in several stable oxidation states. The standard reduction potential for $Tl^{3+}_{(aq)} + 2e^- \rightarrow Tl^+_{(aq)}$ is 1.28 V ($K = 10^{43.3}$) (Kotvly and Sucha, 1985). The high reduction potential points to the fact that Tl(III) has a limited stability field and that Tl(I) should be the dominant species under most environmental conditions, a fact that has been documented in several studies (Bodek et al. 1988; Flegal and Patterson, 1985; Vink, 1993). Assuming thermodynamic equilibrium, the Tl(III)/Tl(I) ratio can be calculated by using the Nerst equation, $\varepsilon = \varepsilon^0 - 2.303RT/nF \log[Tl^+]/[Tl^{3+}]$; this

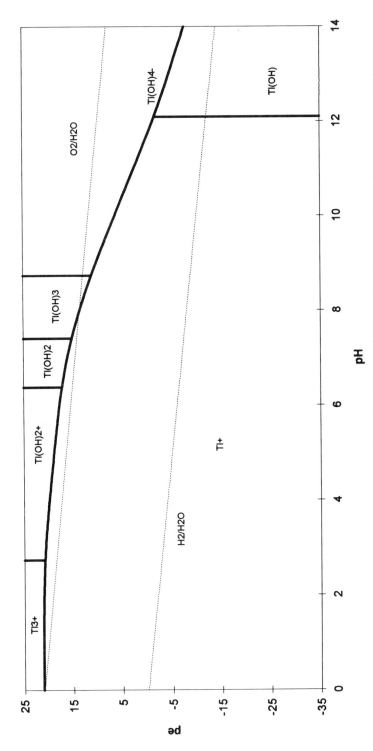

Figure 8. Oxidation–reduction (pe–pH) predominance diagram for Tl(I) and Tl(III) species without considering the precipitation of Tl(OH)$_3$.

40

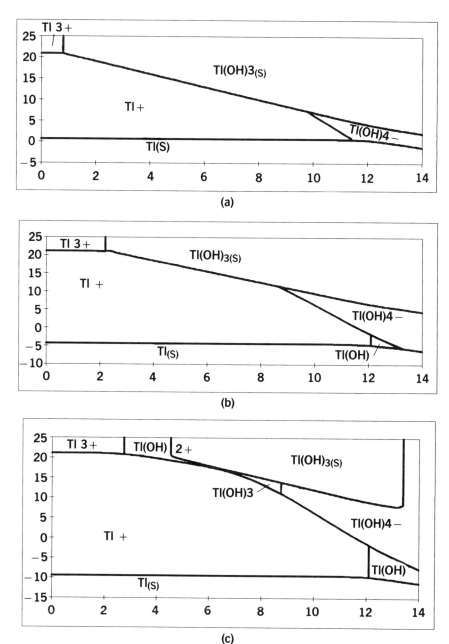

Figure 9. Oxidation–reduction (pe–pH) predominance diagram for Tl–Tl(I)–Tl(III) species. (*a*) Assuming total dissolved Tl = 1 × 10^{-5} *M;* (*b*) total dissolved Tl = 1 × 10^{-10} *M;* (*c*) total dissolved Tl = 1 × 10^{-15} *M.*

expression can be simplified to $\varepsilon = 1.28 - 0.29 \log[Tl^+]/[Tl^{3+}]$. The pe–pH diagram (Fig. 8) for dissolved thallium can be generated by using the hydrolysis constants for Tl(I) and Tl(III) and the equilibrium constant for $Tl^{3+}_{(aq)} + 2e^- \rightarrow Tl^+_{(aq)}$, $K = 10^{43.3}$. The suggestion is that the Tl^+ ion is the most thermodynamically stable species in the aqueous environment and that $Tl(OH)_4^-$ should occur in highly oxidizing and basic environments. Figure 9 includes the formation of solid $Tl(OH)_3$ and elemental Tl in the stability diagram. The calculated fields of dominance suggest that $Tl(OH)_4^-$ is likely to predominate at current Tl concentrations ($<10^{-10}$ M) in the Great Lakes if the pH of waters is higher than 9.0.

Since the Tl(III)–Tl(I) standard potential (1.28 V) is much higher than those of AsO_4^{3-}/AsO_3^{3-} (0.56 V), Fe^{3+}/Fe^{2+} (0.77 V), and SO_4^{2-}/SO_3^{2-} (0.17 V), the reactions between Tl(III) and AsO_3^{3-}, Fe^{2+}, and SO_3^{2-} may occur spontaneously. For example, the equilibrium constant for the reaction $Tl^{3+}_{(aq)} + H_2SO_{3(aq)} + H_2O$ $Tl^+_{(aq)} + 4H^+_{(aq)} + SO_4^{2-}_{(aq)}$, ($K = 3.5 \times 10^{37}$) suggests that Tl(III) should be a powerful oxidant for dissolved SO_2 in water. The role of Tl(III) as an environmental oxidant has not been properly evaluated.

REFERENCES

Baes, C. F., and Mesmer, R. E. (1976). *The Hydrolysis of Cations.* Wiley, New York.

Bidoglio, G., Gibson, P. N., O'Gorman, M., and Roberts, K. J. (1993). X-ray absorption spectroscopy investigation of surface redox transformations of thallium and chromium on colloidal mineral oxides. *Geochim. Cosmochim. Acta,* **57**, 2389–2394.

Bodek, I., Lyman, W. J., Reehl, W. F., and Rosenblatt, D. H. (1988). *Environmental Inorganic Chemistry.* Pergamon, New York.

Cheam, V., Lechner, J. Desrosiers, R., Sekerka, I., Lawson, G., and Mudroch, A. (1995). Dissolved and total thallium in Great Lakes waters. *J. Great Lakes Res.,* **31**, 384–394.

Cronan, D. S. (1976). Manganese nodules and other ferro-manganese oxide deposit. In *Chemical Oceanography,* Vol. 5, Riley, J. P., and Chester, R. (eds.). Academic, New York.

Cotton, R. A., and Wilkinson, G. (1972). *Advance Inorganic Chemistry,* 3rd ed. Interscience, New York.

Dean, J. A. (1985). *Lange's Handbook of Chemistry.* McGraw-Hill, New York.

Dzombak, D. A., and Morel, F. M. M. (1990). *Surface Complexation Modeling.* Wiley, New York.

Flegal, A. R., and Patterson, C. C. (1985). Thallium concentrations in seawater. *Mar. Chem.,* **15**, 327–331.

Kotvly, S., and Sucha, L. (1985). *Handbook of Chemical Equilibrium in Analytical Chemistry.* Wiley, New York.

Lee, A. G. (1971). *The Chemistry of Thallium.* Elsevier, Amsterdam.

Magorrian, T. R., Wood, K. G., Michalovic, J. G., Oek, S. L., and Van Lier, M. M. (1974). *Water Pollution by Thallium and Related Metals.* National Technical Information Service (NTIS) Publication No. PB253 333, Springfield, VA.

Muller, B., and, Sigg L. (1990). Interaction of trace metals with natural particle surface: Comparison between adsorption experiments and field measurements. *Aquatic Sci.,* **52**, 75–92.

Riley J. P., and Siddqui S. A. (1986). The determination of thallium in sediments and natural waters. *Anal. Chim. Acta,* **181,** 117–123.

Schiewer, S., and Volesky, B. (1996). Modeling multi-metal ion exchange in biosorption. *Environ. Sci. Technol.,* **30,** 2921–2927.

Vink, B. W. (1993). The behavior of thallium in the (sub)surface environment in terms of Eh and pH. *Chem. Geol.,* **19,** 119–123.

4

THALLIUM IN THE (SUB)SURFACE ENVIRONMENT: ITS MOBILITY IN TERMS OF Eh AND pH

Bernhard W. Vink

*Department of Geology, University of Botswana,
Private Bag 0022, Gaborone, Botswana*

Thallium in the Environment, Edited by Jerome O. Nriagu.
ISBN 0-471-17755-5. © 1998 John Wiley & Sons, Inc.

4.1. INTRODUCTION

In natural materials thallium occurs mostly in the monovalent and/or trivalent form, Tl^+ and/or Tl^{3+}, respectively. The average thallium content of Earth's crust is very low, being measured in part per million levels or even less. It appears to be slightly enriched in continental granitoid rocks, but even in those rocks the thallium content will rarely exceed 6 ppm. There is certainly a rather strong enrichment of thallium in residual magmatic rocks such as pegmatites, for which thallium contents of up to 150 ppm have been recorded (De Albuquerque and Shaw, 1970).

Most of the thallium in Earth's crust occurs in the crystal lattices of rock-forming minerals, by isomorphic substitution with the K^+ ion. As such it may be enriched in minerals such as biotite (up to 15 ppm) and K feldspar (up to 6 ppm). In pegmatites these minerals may contain much more thallium, in micas up to 100 ppm, and in K feldspars occasionally up to several hundred parts per million.

Together with this information, distinct thallium minerals are listed by De Albuquerque and Shaw, (1970). The following 10 thallium minerals are mentioned, and these are all extremely rare and make probably only a minor contribution to the total thallium content of Earth's crust:

Lorandite, $TlAsS_2$
Picotpaulite, $TlFe_2S_3$
Chalkothallite, Cu_3TlS_2
Vrbaite, $Hg_3Tl_4As_8Sb_2S_{20}$
Hutchinsonite, $(Pb, Tl)_2(Cu, Ag)AS_5S_{10}$
Bukovite, $Cu_{3+a}Tl_2FeS_{4-a}$
Wallisite, $PbTlCuAs_2S_5$
Hatchite, $PbTlAgAs_2S_5$
Crookesite, $(Cu,Tl,Ag)_2Se$
Avicennite, Tl_2O_3

Thus eight of these minerals are sulfides, one is a selenide, and one is an oxide. Except for the oxide avicennite, thermodynamic data (Gibbs free energies of formation and absolute entropies) are not known, and for that reason the avicennite is the only natural thallium mineral that can be included in the forthcoming calculations and considerations. However, considering the extreme scarcity of distinct natural thallium minerals, this is of minor importance. As a general trend, the eight sulfide minerals and the selenide mineral will be stable only under rather strongly reducing conditions, be it acidic or alkaline, but more likely alkaline since the synthetic simple sulfide Tl_2S can only exist under extreme alkaline conditions.

In the surface and subsurface environments, thallium will be released by weathering processes, by breakdown of thallium-bearing minerals, be it the distinct minerals listed or minerals that have thallium locked up in their crystal lattices. It is here that considerations concerning the behavior of thallium in such environments are important: Eh–pH diagrams may be of help to explain and/or predict the behavior of thallium under more or less atmospheric conditions.

4.2. Eh–pH DIAGRAMS AT 1 ATM TOTAL PRESSURE AND 25°C

Brookins (1988, p. 52) published the first Eh–pH diagram for thallium. This diagram is reproduced here as Figure 1.

In his diagram, the monovalent thallium ionic species Tl^+ occupies the reducing area of Eh–pH space, whereas transitional and oxidizing areas are occupied by the thallium oxides Tl_2O, Tl_2O_3 (avicennite), and Tl_2O_4. Thallium sulfide, Tl_2S, can be found to be stable only under extreme alkaline reducing conditions. The diagram configuration suggests thallium to be immobile under transitional and oxidizing conditions, be it acidic or alkaline, being precipitated as one or more of the thallium oxide phases.

Unfortunately, with the exception of the Tl_2S phase, Brookins's thallium diagram cannot be considered correct. For instance, the transition from Tl^+ to Tl_2O cannot possibly be a line with a negative slope in the diagram. The reaction:

$$2Tl^+ + H_2O \leftrightarrows Tl_2O + 2H^+ \qquad (1)$$

is Eh-independent and as such must be a vertical line in the diagram. Using the same data as Brookins (1988), this Tl^+/Tl_2O transition is found to be at pH = 21.55, that is, far outside "normal" Eh–pH space.

The diagram was revised and corrected by Vink (1993). The main and rather unusual aspect is that most of the calculations result in phase transitions outside the normal Eh–pH space, at pH values far in excess of 14 (Fig. 2; after Vink, 1993). Figure 3 (also after Vink, 1993) gives the final diagram for 25°C and 1 atm total pressure, within the normal pH limits between 0 and 14 and Eh limits between −0.8 and +1.4 V.

In this new diagram, the monovalent thallium ionic species Tl^+ now occupies almost all Eh–pH space. Only under very alkaline and oxidizing conditions do thallium oxides feature (Tl_2O_3, avicennite, and Tl_2O_4, no natural mineral), and under very alkaline reducing conditions the simple sulfide Tl_2S (no natural mineral).

The diagram of Figure 3 was calculated for a thallium activity of 10^{-8} M, equivalent to approximately 2 ppb, and contoured for activity 10^{-6} M, equiva-

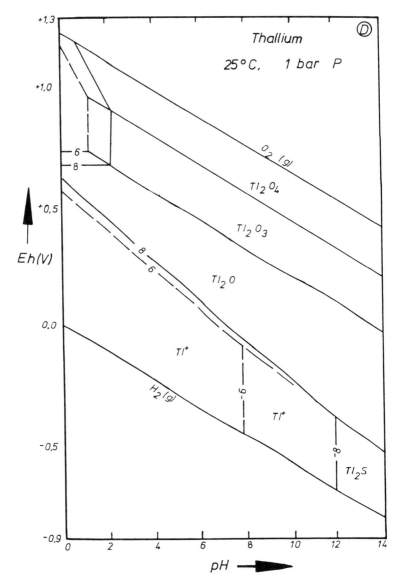

Figure 1. Eh–pH diagram for part of the system Tl–S–O–H, at 25°C and 1 atm total pressure. The assumed activities for dissolved species are: Tl $= 10^{-8}$ and 10^{-6}, S $= 10^{-3}$. After Brookins (1988, Fig. 23, p. 52).

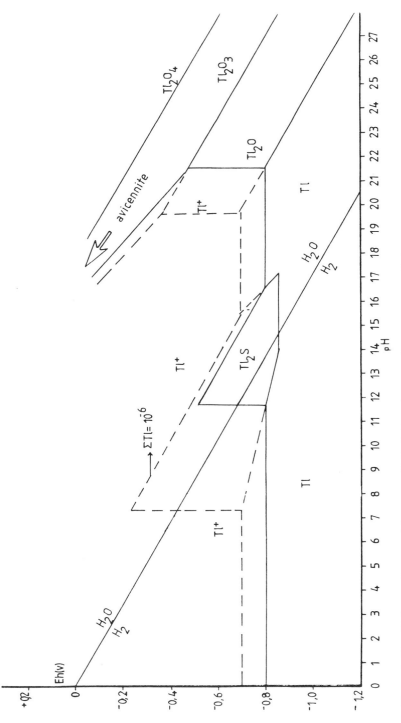

Figure 2. Eh–pH diagram for part of the system Tl–S–O–H, at 1 atm total pressure; "unrealistic" areas for pH values varying from 0 to 28, at negative Eh values. Activity of total dissolved thallium species: 10^{-8} m (contoured for 10^{-6} m). Activity of total dissolved sulfur species: 10^{-3} m.

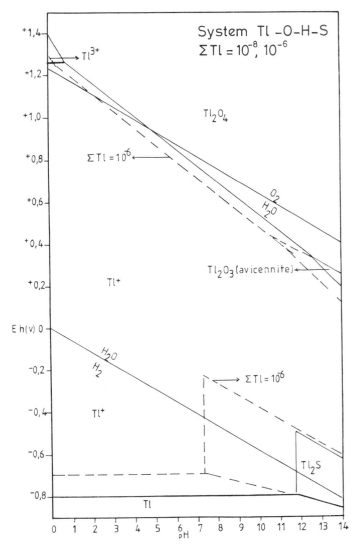

Figure 3. Eh–pH diagram for part of the system Tl–S–O–H, at 1 atm total pressure; "normal" areas. Activity for total dissolved thallium species: 10^{-8} m (contoured for 10^{-6} m). Activity of total dissolved sulfur species: 10^{-3}.

lent to 204 ppb. Increase in activities results in a minor increase of the oxide fields and a substantial increase of the Tl_2S field.

The main outcome of Figure 3 is that thallium must be considered to be a very mobile element: Under almost all natural conditions, Tl^+ is the main phase in the diagram, and only under extreme conditions will the ionic species precipitate as oxide or as sulfide. This is entirely in line with the empirical observation of De Albuquerque and Shaw (1970) who report that thallium tends to disperse readily during oxidation of thallium-bearing sulfide deposits.

4.3. TEMPERATURE DEPENDENCE OF THE DIAGRAM

Mainly in the interest of environmental issues, it may be useful to establish the temperature dependence of the diagram, between the normal 25 and 100°C.

4.3.1. Calculation Method

The Gibbs free energy of reaction at temperature T, of any one of the reactions concerned, is calculated, using the standard thermodynamic equation:

$$\Delta G°_T(r) = \Delta G°_{298}(r) - \Delta S°_{298}(r)(T - 298) \qquad (2)$$

in which $\Delta G°$ 298 (r) is the Gibbs free energy of the reaction at 298 K, and $\Delta S°_{298}(r)$ the entropy of reaction at 298 K.

This calculation is followed by the standard application of Nernst's law:

$$Eh = \Delta G°_T(r)/nF + RT/nF \ln K \qquad (3)$$

in which F is the Faraday constant (96.48 kJ/V gram equivalent), R the gas constant (0.0083 kJ/deg), n is the number of electrons in the reaction equation, T is the absolute temperature in Kelvin, ln is the naperian logarithm, and K is the equilibrium constant. This results in the Eh–pH equation of the reaction concerned. It should be noted that the slope of the lines in the diagram changes with temperature, being governed by the term RT/F, including the slopes of the water stability boundaries.

In these calculations, the thallium activity was taken again as 10^{-8} M (~2 ppb), and the diagram was then contoured for activity 10^{-5} M (~2 ppm). Diagrams were calculated for 35 and 100°C and these are shown in Figures 4 and 5. They also include the areas outside normal Eh–pH space, that is, areas for pH values between 14 and 24, where the thallium oxide stability relations are to be calculated and established ("unrealistic" areas).

4.3.2. Thermodynamic Data

The thermodynamic data used in the calculations of the diagrams of Figures 2–5 are listed in Table 1.

As far as possible, and in fact the majority of data, were taken from Wagman et al. (1982). Only in cases where Wagman et al. (1982) do not list the required

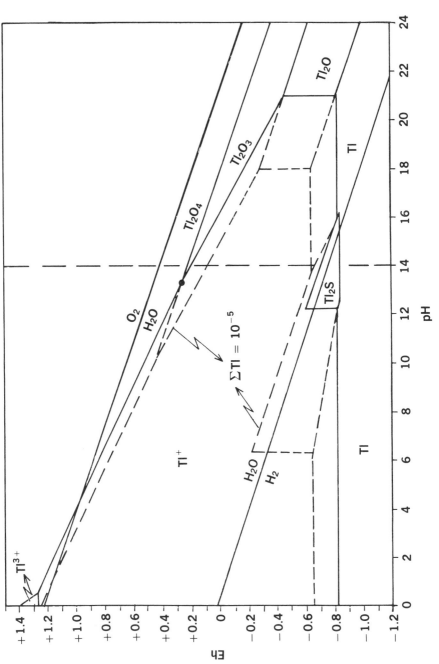

Figure 4. Eh–pH diagram for part of the system Tl–S–O–H, at 35°C and 1 atm total pressure, both "unrealistic" and "normal" areas. Activity for total dissolved thallium species: 10^{-8} m (contoured for 10^{-5} m). Activity of total dissolved sulfur species: 10^{-3} m.

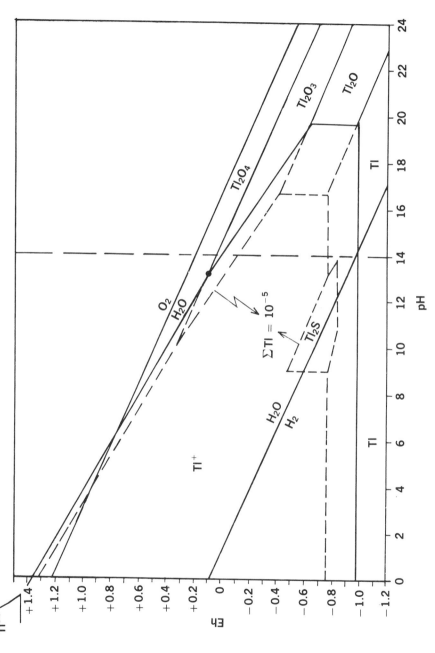

Figure 5. Eh–pH diagram of the system Tl–S–O–H, at 100°C and 1 atm total pressure, both "unrealistic" and "normal" areas. Activity for total dissolved thallium species: 10^{-8} m (contoured for 10^{-5} m). Activity of total dissolved sulfur species: 10^{-3} m.

Table 1   Thermodynamic Data. $\Delta G^\circ_{298}{}^a$ and $S^\circ_{298}{}^b$

Species (State)	ΔG°_{298}	Reference	S°_{298}	Reference
Tl(c)	0	—	+64.3	SGTE, 1994
Tl^+(aq)	−32.38	Wagman et al., 1982	+127.2	Weast, 1968–69
Tl^{3+}(aq)	+209.20	Weast, 1968–69	−443.5	Weast, 1968–69
Tl_2O(c)	−147.28	Wagman et al., 1982	+134.1	SGTE, 1994
Tl_2O_3(avicennite)	−311.71	Wagman et al., 1982	+160.0	SGTE, 1994
Tl_2O_4(c)	−347.19	Wagman et al., 1982	+165.2	Approximated
Tl_2S(c)	−93.68	Wagman et al., 1982	+159.0	SGTE, 1994
TlCl(c)	−204.97	Weast, 1968–69	+108.4	Weast, 1996–69
H_2O(l)	−237.10	Wagman et al., 1982	+69.9	Wagman et al., 1982
HS^-(aq)	+12.09	Wagman et al., 1982	+62.8	Wagman et al., 1982
S^{2-}(aq)	+85.81	Wagman et al., 1982	−14.6	Wagman et al., 1982
SO^{2-}_4(aq)	−744.54	Wagman et al., 1982	+20.1	Wagman et al., 1982
NaCl(c)	−384.10	Wagman et al., 1982	+72.1	Wagman et al., 1982
Cl^-(aq)	−131.20	Wagman et al., 1982	+56.5	Wagman et al., 1982
Na^+(aq)	−261.90	Wagman et al., 1982	+59.0	Wagman et al., 1982

[a] Values in kilojoules per gram formula weight.
[b] Values in joules per degree per gram formula weight.

data were other sources used. Even so, one datum could not be found anywhere, that is, the absolute entropy value for the oxide Tl_2O_4.

This value was approximated as follows. If the transitions Tl/Tl_2O and Tl_2O/Tl_2O_3 are calculated for 100°C, then both lines, compared with 25°C, move upward by 0.025 Eh units. It is logical and acceptable then that also the boundary Tl_2O_3/Tl_2O_4, which has the same negative slope as the other two boundaries, moves up by the same amount. Taking the position of that line thus obtained for 100°C, one can work backward toward establishing the $\Delta G^\circ_{373}(r)$ value and from there the $\Delta S^\circ_{298}(r)$ value, leading eventually to the S°_{298} (Tl_2O_4) value, which was thus approximated as +165.2 J/deg.

4.3.3.   Diagrams at Elevated Temperatures

At increasing temperatures, the boundary Tl/Tl^+ moves gradually down, whereas the boundary Tl/Tl_2S in the S^{2-}–sulfur ionic species field moves gradually up. It is a simple algebraic calculation to find that at 38°C, using the data of Table 1, these boundaries are level and that as a result the Tl_2S field disappears at temperatures in excess of 38°C. But this is only true for a thallium activity of 10^{-8} M: At higher activities, Tl_2S will still retain a stability field.

The diagram of Figure 4 at 35°C shows that the above-mentioned boundaries are almost level (the main reason to include this diagram in this contribution). Otherwise the diagram for 35°C is identical to that for 25°C.

The diagram for 100°C (Fig. 5) shows that Tl_2S indeed has disappeared at Tl^+ activity 10^{-8} M, but this field reappears at strongly increased activities for which a contoured example is given for 10^{-5} M (roughly equivalent to 2 ppm).

A comparison between the diagrams at 25 and 100°C reveals that the oxide fields of Tl_2O_3 and Tl_2O_4 have slightly decreased, whereas the Tl_2S field has entirely disappeared, for $Tl^+ = 10^{-8}$ M. Therefore an increase in temperature only increases the stability field of Tl^+, or in other words an increase in temperature enhances the mobility of thallium.

This would indicate already that in addressing problems concerning treatment of thallium-bearing wastewater or thallium-polluted groundwater elevated temperatures should not be considered because that only increases the mobility of thallium.

4.4. SUGGESTIONS FOR POLLUTION PROBLEMS

4.4.1. Wastewater Treatment and Sludge Disposal

Within the realm of chemistry, Eh–pH diagrams resort under the discipline of the chemistry of very dilute aqueous solutions. Thallium is a very toxic metal and as such concentration levels of several parts per million may already be considered as pollution. Although the boundaries "polluted" and "nonpolluted" waters may be arbitrary, it is taken here that the thallium activity of 10^{-8} M, as used in the construction of the diagrams of Figures 2–5 can be considered normal (although perhaps relatively high), and the contouring for 10^{-6} or 10^{-5} M (0.2–2 ppm) can be considered to be a pollution concentration. In view of the high toxicity of thallium, it will generally be taken as a matter of priority to clean polluted waters of this toxic material by precipitating the Tl^+ ion in any insoluble or barely soluble thallium compound. The Eh–pH diagrams as shown in Figures 1 to 5 are here of little help: They only show the extreme mobility of thallium under virtually all natural surface and/or (sub)surface conditions. One may try, of course, to create a strongly alkaline environment by adding certain chemicals, followed by addition of a strongly oxidizing agent in order to precipitate the Tl^+ ion as Tl_2O_3 and/or Tl_2O_4. But it would appear that an easier and probably less harmful procedure could be attempted, at least from a theoretical point of view: Experimental work should be carried out to confirm theoretical considerations.

A suitable medium to precipitate the Tl^+ ions may be found in TlCl. Although its solubility product K_s is not tabulated, this can be readily approximated by using the thermodynamical data of Table 1:

$$TlCl(c) \leftrightharpoons Tl^+(aq) + Cl^-(aq) \qquad (4)$$

$$\Delta G^\circ_{298}(r) = +41.39 = -RT \ln K_s$$

which results in

$$K_s = [Tl^+][Cl^-] = 10^{7.26}$$

This indicates that thallium chloride is a rather insoluble compound with a suitably low solubility product.

If a concentration of 2 ppm thallium in wastewater is taken as an example, or in other words an activity of approximately 10^{-5} M, then Cl^- ions could be added in order to create a Cl^- activity in excess of $10^{-2.26}$ M, in order to precipitate thallium chloride, provided that the Cl^- ions do not interact with other possible chemical substances in the wastewater.

Addition of the chloride ionic species could be done in several ways, and hydrochloric acid might be an obvious candidate. But from a merely theoretical point of view, it would appear that another less harmful alternative can be found in the form of sodium chloride. Assuming complete ionization, a similar calculation for a saturated solution of NaCl at 25°C shows

$$NaCl(c) \leftrightharpoons Na^+(aq) + Cl^-(aq) \qquad (5)$$

$$\Delta G^\circ_{298}(r) = -9.00 = RT \ln K_s$$

which results in

$$K_s = [Na^+][Cl^-] = 10^{+1.58}$$

In a saturated salt solution then

$$[Na^+] = [Cl^-] = 10^{+0.79} \text{ M}$$

This is in very good agreement with the empirical value for a saturated salt solution, which is $10^{+0.81}$ M. This chloride concentration is far in excess of the value of $10^{-2.26}$ M needed, and the 2 ppm thallium as Tl^+ ions should readily precipitate as TlCl.

As a result of the nature of the solubility product, the lower the thallium concentration in the wastewater, the higher the chloride activity is needed to compensate for this. At thallium levels down to 2 ppb ($= \sim 10^{-8}$ M): $[Tl^+] =$

10^{-8}, then $[Cl^-]$ must be $10^{+0.74}$ M, and the saturated salt solution should still, but only just, clean the wastewater from the Tl^+ ions.

It should be emphasized again that this conclusion is based only on strictly theoretical considerations concerning water polluted by Tl^+ and that experimental work should be carried out to confirm this. But it would appear that addition of solid sodium chloride should clean the wastewater from Tl^+, down to levels as low as 2 ppb. In light of the considerations on the temperature dependence of the Eh–pH diagram, it would be advisable to carry out such experiments at room temperature or only slightly higher temperature because increase in temperature only increases the mobility of thallium.

4.4.2. Polluted Groundwater

Groundwater containing only a few parts per million thallium may be considered a serious hazard, and unfortunately in this case it is more difficult to make useful suggestions to tackle such a problem.

A clear understanding and knowledge of the local conditions where thallium pollution took place is needed in the first place: Factors such as the relief of the polluted area, the depth of the water table, and velocity of groundwater movement must be known before any action can be undertaken. More important probably is the chemical and mineralogical composition of the soil or any other weathered surface material.

If an area is concerned that is covered by soils and/or sand, simple hand auger drill holes may be made, if the water table is less than 5 m deep. Grid density will depend entirely on local conditions, mainly groundwater flow. If hand augering penetrates the water table, the drill holes may be filled with a suitable neutralizer. Most chemicals that can remove thallium from a very dilute groundwater solution will not be suitable since they may make the pollution problem even worse, ruling out, for instance, hydrochloric acid.

Substantial amounts of thallium in soils may be adsorbed by clay minerals, although it is difficult to quantify this entity in case of a pollution problem. Shaw (1952) emphasizes the relatively high thallium content of argillaceous sediments, averaging 0.69 ppm, as compared with an apparent thallium shortage in ocean waters residues. Also sandstones and other arenaceous sediments may be enriched in thallium, averaging 0.82 ppm (Shaw, 1957; De Albuquerque and Shaw, 1970), although it appears to be unknown in which form it occurs, that is, locked up in the crystal lattices of some minerals or adsorbed by some of the mineral components.

Adsorption phenomena cannot be included in Eh–pH diagrams, and for that reason these diagrams cannot be of much help. The only suggestion that may be given here would be to attempt the same method suggested in Section 4.4.1 that is, addition of a saturated salt solution in order to precipitate thallium chloride. This, of course, may result in a temporarily strongly increased salinity of the groundwater, but if the thallium pollution is very serious indeed, this would appear to be preferable to thallium poisoning.

ACKNOWLEDGMENTS

Dr. W. Ddamba of the Chemistry Department of the University of Botswana is gratefully acknowledged for the fruitful discussions concerning thallium precipitation, and Dr. M. Jacobs of the Institute of Earth Sciences of the Utrecht State University in the Netherlands is acknowledged for his assistance in supplying some of the crucial thermodynamic data.

REFERENCES

Brookins, D. G. (1988). *Eh–pH Diagrams for Geochemistry.* Springer-Verlag, Berlin.

De Albuquerque, C. A. R., and Shaw, D. M. (1970). Thallium. In *Handbook of Geochemistry, Sect. 81A–810,* Wedepohl, K. H. (ed.). Springer-Verlag, Berlin.

SGTE (Scientific Group Thermodata Europe) (1994). Substance Database SUB94. Grenoble, France.

Shaw, D. M. (1952). The geochemistry of thallium. *Geochim. Cosmochim. Acta,* **2,** 118–154.

Shaw, D. M. (1957). The geochemistry of gallium, indium, thallium—a review. *Phys. Chem. Earth,* **2,** 164.

Vink, B. W. (1993). The behaviour of Thallium in the (sub)surface environment in terms of Eh and pH. *Chem. Geol.,* **109,** 119–123.

Wagman, D. D., Evans, W. H., Parker, V. B., Schumm, R. H., Halow, I., Bailey, S. M., Churney, K. L., and Buttall, R. L. (1982). The NBS Tables of chemical thermodynamic properties. Selected values for inorganic and C1 and C2 organic substances in SI units. *J. Phys. Chem. Ref. Data,* **11,** Suppl. 2, 392.

Weast, R. C. (ed.), (1968–69). *Handbook of Chemistry and Physics.* Chemical Rubber Co., Cleveland.

5

THALLIUM IN AGRICULTURAL PRACTICE

Manfred Sager

Bundesamt und Forschungszentrum für Landwirtschaft,
Institut für Agrarökologie, Spargelfeldstraße 191, A-1226 Wien,
Austria

Thallium in the Environment, Edited by Jerome O. Nriagu.
ISBN 0-471-17755-5. © 1998 John Wiley & Sons, Inc.

5.1. SOILS AND SEDIMENTS

5.1.1. Geogenic Background Level

In silicates, the geochemical behavior of thallium is dominated by the isomorphism of anhydrous monovalent Tl with Rb (Rb:Tl is about 300:1) and K. Contrary to most of the other heavy trace elements, the Tl content is clearly higher in acid rocks than in basic ones. Granites were found to contain 2.7–4.0 $\mu g/g$ (Butler, 1962) or 0.35–3.60 $\mu g/g$, respectively (Siedner, 1968), but in basalts only 0.021–0.058 $\mu g/g$ were encountered (Heinrichs, 1975). The high solubility and low complexation ability of Tl^+ leads to low contents in carbonates and clay minerals (limestone: 0.10–0.14 $\mu g/g$; after Heinrichs, 1975). In sulfides, Tl can be immobilized and oxidized to the trivalent state by elemental sulfur, -resulting in levels above its average abundance of the earth's crust. The major part of Tl within the earth's crust is bound to pyrite; proper Tl minerals are rarely found.

A screening program in the Province of Upper Austria clearly revealed that thallium contents in soils strongly depended on the original geological material (Table 1). In total 460 soil samples were taken from areas of agricultural use, derived from all geological formations. The samples were sieved <2 mm, milled, and 1 g was extracted with 6 mL HNO_3 for 5 h under reflux (Hofer et al., 1990). Range and mean were not significantly different with respect to usage as arable soil or grassland. Of these 28 soil profiles showed constant thallium contents or slightly higher contents in subsoils. No thallium emission source was detected in the province. Higher thallium levels in the crystalline is probably due to enrichment in K feldspars. No significant correla-

Table 1   Tl Contents of Upper Soils (0–20 cm) in Upper Austria

Geological Formation	Number	Range	Median
Molasse	64	0.076–0.288	0.180
Alluvial terraces	134	0.080–0.325	0.190
Sandstone	50	0.172–0.410	0.243
Alpine limestone	34	0.099–0.754	0.317
Bohemian crystalline	122	0.141–0.91	0.484

Source: Hofer et al. (1990).

tions between pH, clay size fraction, humus, and thallium were found (Hofer et al., 1990).

In China, within a screening program for Tl contents in arable soils, 853 samples from 34 provinces were investigated. Thallium contents ranged from 0.292 to 1.172 μg/g, and were Gaussian-like distributed, with a mean of 0.580 μg/g. In steppe soil and desert soil, the Tl amounts were lowest. It increased in dark brown forest soil of 7.8% organic substance and became particularly high in podzolic soil and black soil with more organic substance, more physical clay, and not alkaline pH (pH 6.8). The group of alpine soil contained the highest amount of Tl, due to late formation and weaker leaching of the soil. Therefore, in geographical terms, it decreased from east to west, and in the provinces and cities south of the Chanjiang River, the values for Tl were clearly higher than those for north and northwest China (Qi et al., 1992).

Limestone soil may contain high amounts of Tl because 627 m³ of limestone form only 1 m³ of weathered substance in which Tl is concentrated. Contrary to the findings in Europe, in China Tl showed one of the most positive correlations with organic substance contents (Qi Wenqi et al., 1992).

In presumably noncontaminated soil samples from Baden Württemberg (Germany), a mean thallium contents of 0.4 μg/g was found (Scholl, 1980). In vineyard soils at Ingelheim, Rhine, Germany, intensely used for centuries, the mean of 0–20 cm soil was 0.26 μg/g, and of 20–40 cm soil only 0.16 μg/g (Eschnauer et al., 1984). In France, 200 soil samples from rural areas and 50 soil samples from anthropogenic areas were screened for their Tl contents. The samples were sieved <2 mm, air dried, milled, and digested with HNO_3/H_2O_2. A range of 0.1–50 μg/g Tl was found. Soil samples with high geogenic Tl contents were of clay type, developed on hard Sinemurian limestone, and contained also elevated levels of Pb, Ag, Zn, Ni, and Cd, often located inside goethite and Mn oxides shots (Tremel et al., 1995).

5.1.2. Soil Contamination

In Germany, Tl has been nominated among nine priority inorganic pollutants for soil in 1985 (Severin et al., 1990). Thallium contamination pathways are atmospheric emission, as well as weathering and washout of residues from mines or smelters.

Emission sources of thallium have been mainly nonferrous mines, cement plants, and coal burning power plants (Kaplan et al., 1990). As Tl is volatile both under reducing and oxidizing conditions, as chloride, nitrate, oxide, or element, it is enriched in the flue dust of smelters, cement plants, and the like. Similarly, Tl from coal is volatilized during combustion processes; crude oil contains lower Tl amounts. Like other atmospherically transported contaminants, Tl is enriched in upper soil layers in case of emission from atmospheric sources. Even in the surroundings of brickwork, up to fivefold increase of thallium contents of soil and grass was detected (Brumsack, 1977) because of volatilization during calcination and subsequent deposition.

In soils from Germany, HNO_3^- extractable fractions ranged from 0.2 to 4.0 mg/kg. Within 14 years, the mobility of Tl in these contaminated soils dropped down to 10–15% of the HNO_3^- extractable fraction, perhaps because of a soil pH > 7. But still more transfer into plants occurred than from the same level present in geogenic fractions (Crössmann, 1994).

The Tl in the flue dust did not emanate from the cement raw material itself, but from $Fe_2O_3^-$ containing additives, obtained as a residue from roasting of pyrite (Scholl, 1980; see Table 2). Investigations on soil cores (pH about 4.0) and trace element fluxes within a beech forest in central Germany showed thallium and other trace elements to decrease from the top to deeper layers. Considering the input from atmospheric deposition and output from run-off waters, a positive input/output balance was obtained, which was not balanced by increasing biomass. Thus, this forest ecosystem acted as a filter for substances transported via the atmosphere (Heinrichs and Mayer, 1977).

Even higher levels were found at sites of historic mining activities in Northern Badenia, Germany (Scholl, 1980; Puchelt and Walk, 1980). Whereas soils in the neighborhood of former mining activities contained Tl/Cd/Pb/Bi only at background levels, sites influenced from mining showed enrichment of Tl (3–35-fold), Cd and Pb, but Bi remained low. In Germany, 1 mg/kg Tl in soils has been established as tolerance level for agricultural use (Kloke, 1980).

Table 2   Tl Contents of Soils in Baden, Württemberg

Sample Type	Number	Mean	Range
Soils noncontaminated	478	0.4	0.1–2.2
Soils near cement plants	575	1.4	0.1–15
Soils near old mines	221	14.7	0.1–73
Cement raw material	11	0.45	0.13–2.6
Flue dust from cement plants	21	428	42–2370

After Scholl (1980).

5.1.3. Mobile Fractions

Contrary to Zn, Pb, Cd, and Cu in soils, extraction with aqua regia (5 g soil/ 20 mL; 2 h reflux) yielded only 28–74% of total contents (Lukaszewski and Zembrzuski, 1992). In soil samples from Bulgaria, aqua regia and conc. HCl leached about half of the amount obtained with HNO_3/HF. Only 6% of total was found exchangeable with NH_4Cl, pH = 7, or with KCl (Tsakovski et al., 1994).

In a selective leaching sequence, modified after Tessier et al. (1979), in presumably nonpolluted river sediments, large amounts were found leachable in oxalate and hot HNO_3, which indicates preferential affinities toward Fe hydroxides, sulfides, and silicates. The sediments quantitatively adsorbed Tl from solution at ambient pH overnight, and released it shortly afterward into the exchangeable fraction (1 M ammonium acetate, pH = 7 or 1 M NH_4Cl, pH7). In an alternative sequence, appreciable amounts were found alkali mobile, especially in a calcareous clay sample, contrary to most other trace elements (except As, Zn) (Sager, 1992b).

At a contaminated site in Germany with more than 50% clay content (mainly illite), large fractions of thallium were exchangeable versus ammonium acetate, pH = 7, and residual. The contamination with thallium, as well as the exchangeable proportion strongly increased with decreasing grain size, thus demonstrating the effect of active surfaces (Lehn and Schoer 1987). From cement, leaching of Tl with tap water and CO_2-saturated water was negligible (Sprung et al., 1994).

In the sediment of the Lenne River, Germany, which was polluted with 21 mg/kg Tl from a smelter ("Germaniahütte"), Tl was significantly mobilizable only at pH < 2, whereas complexing agents like nitriloacetic acid (NTA) and bis(2-ethylhexyl) phosphoric acid (HEDP), were of no influence on mobilization (Gunther et al., 1987). Similarly, in sediments of the Rhine and Elbe Rivers in Germany only 43–86% of Tl were found after extraction with aqua regia (Waidmann et al., 1992); the extractability with aqua regia was smaller at low concentration levels.

In fine sediment samples from Lake Constance, 1 M HNO_3 leached about 60% Tl of total (Waidmann et al., 1994). In sediments from the Elbe River (Germany), aqua-regia-extractable Tl was not significantly different from the amount extractable with 1 M HNO_3. The total amount, obtained after dissolution with HNO_3/HF, was about 0.3–0.4 mg/kg higher for all samples (Waidmann, et al., 1992).

5.1.4. Soil to Plant Transfer

The transfer from soil to plants generally depends on species-specific properties, as well as soil properties, such as clay size and humus contents, cation exchange capacity, and pH. The thallium uptake by green plants, above all rape (*Brassica napus*), correlated with the exchangeable soil fraction; minor amounts were assumed to be transferred to the plant from the hydroxylamine-reducible fraction (Lehn and Schoer, 1987).

Addition of Tl-containing flue dust to increase the HNO_3^- leachable contents from 0.4 to 0.9 mg/kg did not significantly influence the crop yields for green rape, bush beans, and rye grass (Makridis and Amberger, 1989a). The Tl concentrations in all green plants uniformly increased with increasing Tl available. At pH 5.6, about 50% higher transfers to the plants occurred than at pH 6.2. Green rape strongly accumulated Tl and reached higher concentrations than the substrate soil, even after addition of only 0.05 mg/kg to the soil, whereas the Tl contents of bush beans and rye grass increased only marginally (Makridis and Amberger, 1989a). Rape and cabbage have high Tl- accumulating abilities, and are not recommended to be grown within a Tl-polluted area. Tolerance levels, 1 mg/kg dry weight for rape, and 0.1 mg/kg fresh weight for fruits and vegetables, are suggested by the German Bundesgesundheitsamt (Crössmann, 1994). In vegetables tested for several years on the same polluted soil, the Tl contents declined from year to year, because of the decline of the Tl mobility in the soil (Crössmann, 1994).

5.2. THALLIUM CONTAMINATION SOURCES IN AGRICULTURE

5.2.1 Atmospheric Deposition

Crops and soils are subject to atmospheric deposition, which cannot be influenced by the individual farmer. Atmospheric deposition measurements are widely lacking, but data on flue dusts and fly ash are available, as well as on ambient levels in air. In the Arctic, ice and snow cores may serve as dateable indicators for long-range transported amounts at different periods of time. Ranges for Tl in snow were 0.24–2.2 pg/g (Sturgeon et al., 1993).

In remote areas, such as above the Atlantic Ocean, the concentration of thallium in air is below 0.02 ng/m^3 (Völkening and Heumann, 1990), whereas in the urban area of Genova, Italy, a geometric mean of 15 ng/m^3 has been found (Valerio et al., 1988). In ambient air in the outskirts of Berlin, the 6-h average of Tl contents changed between 0.05 and 1.00 ng/m^3 in the course of a year. The Tl level in exhaust fumes of a natural gas power station was 7.50 ng/m^3, and thus clearly above ambient levels. Airborne particulate matter from the power station and in the ambient air were measured by impaction; maxima at about 2 μm diameter were observed (Tilch et al., 1996). Flue dusts from a zinc mine in Northern Badenia, Germany, contained Tl within 42–670 μg/g, and flue dust from cement plants even up to 2370 μg/g (Scholl, 1980). Emission of these led to significant accumulation in agricultural and garden soils. Thallium in these flue dusts may be highly water soluble (Crössmann, 1994).

Coal combustion is regarded as the most frequently occurring emission source for thallium. Thallium in 106 coal samples, used for generation of electric power in Austria, was found to be within the range <0.1–3.3 μg/g. Within 26 slag and ash samples obtained from the same facilities, Tl contents ranged <0.1–2.41 μg/g. On the average, Tl in the group of slag/fly ash samples was higher (Sager, 1993). Brown coal fired in a 2500-MW power plant in Germany in 1977 had an average content of 0.011 μg/g Tl, the corresponding furnace bottom ash had 0.089 μg/g, and the precipitator ash 0.22 μg/g Tl. This indicates moderately volatile compounds (Heinrichs, 1977). Samples of gypsum obtained in the desulfurization of flue gases are reported to contain thallium in the range <0.005–0.64 mg/kg, which is more than in natural occurring gypsum (0.04–0.13 mg/kg) (Gorbauch et al., 1984); thus, thallium may be partially retained.

5.2.2. Irrigation Water

The need for water supply has been estimated for C_3 plants within the range 500–1000 g/g produced dry substance, and for C_4 plants within the range 260–350 g/g produced dry substance (Schlee, 1986). Thus, if excretion is neglected, and all Tl is incorporated, irrigation water of 1 μg/L would contribute 0.26–1.0 μg/g plant material, which can be regarded as the upper limit for human and farm animal consumption. Ambient levels, however, are far below; see Table 3. Monovalent thallium is not coprecipitated with carbonates but can be, however, adsorbed and coprecipitated on suspended clay minerals (in analogy to K^+), or coprecipitated with Mn hydroxides from low-chloride medium (Magorian et al., 1974; Sager, 1992b). In carbonate-containing alkaline waters, Tl forms labile complexes with humics (O'Shea and Mancy, 1976). In highly oxygenated waters, coprecipitation of trivalent thallium with hydroxides of Fe, Al, Mn, and Mg is possible. On the average, this leads to an accumulation in the sediments of about 50-fold (Berthelot et al., 1980).

In tap water in Poland, sampled 1989–1993, 5–24 ng/L were found; in river waters 5–17 ng/L; and in the Baltic Sea 61 ng/L. The high value found at Bytom (2040 ng/L) is certainly due to nearby Zn ore exploration activities (Lukaszewski et al., 1996).

In the Great Lakes of North America, 87 ± 8% of total Tl has been found to be dissolved. The median contents steadily increased from Lake Superior down to Hamilton Harbor. Across Lake Ontario, 9 stations with a total of 47 depth sampling sites were included. In Lake Ontario, Tl was about evenly distributed across the lake (Cheam et al., 1995). Also for Lake Superior, there was no definite pattern in the vertical profile for Tl, in contrast to dissolved Pb (Cheam et al., 1995). Similarly, Bruland (1983) found a constant Tl level in seawater down to 2500-m depth. At the mouth of the Niagara River, as well as in Hamilton Harbor, higher concentration in the surface were found, possibly because of contamination. The level of soluble thallium present in the sea (e.g., Pacific Ocean, Atlantic Ocean, Irish Sea, Australian Coast) is between 9 and 16 ng/L (Matthews and Riley, 1970). This is remarkably lower than in fresh waters. In natural seawater (pH = 8.1), the oxygen con-

Table 3   Tl Contents in Waters

Sample Type	Country	Waterbody	Concentration	Literature
Tap water	Poland	Var. sources	5.1–24 ng/L[a]	Lukaszewski et al., 1996
River water	Poland	Warta	14 ng/L[a]	Lukaszewski et al., 1996
River water	Poland	Odra	17 ng/L[a]	Lukaszewski et al., 1996
River water	Poland	Pilica	5 ng/L[a]	Lukaszewski et al., 1996
River water	Germany	Var. sources	2–40 ng/L	Heinrichs, 1975
River water	Germany	Rhine (Lobith)	75 ng/L[a]	Cleven and Fokkert, 1994
Lake water	Canada	Lake Superior	1.2 ng/L	Cheam et al., 1995
Lake water	Canada	Lake Ontario	5.7 ng/L	Cheam et al., 1995
Lake water	Canada	Lake Erie	9.4 ng/L	Cheam et al., 1995
Lake water	Canada	Hamilton Harbor	25.7 ng/L	Cheam et al., 1995
Oceans		Pacific, Atlantic	9–16 ng/L	Matthews and Riley 1970
Sea water	Poland	Baltic Sea (Sopot)	61 ng/L	Lukaszewski et al., 1996
Sea water	USSR	Black Sea	82 ng/L	Shevchenko et al., 1977

[a] Unfiltered, + EDTA.

tents is sufficient to oxidize Tl(I) to Tl(III) because formation of chloro-complexes stabilizes the trivalent state. In the Pacific Ocean, 80% of the thallium was found to occur trivalent, and only 20% as the sum of monovalent and alkylthallium-compounds (Batley and Florence, 1975). As trivalent thallium is easily adsorbed and coprecipitated, it continously moves down to the sediments.

Purification of effluents from metallurgical plants has been proposed by sorption on complex Fe(II) cyanides bound to silica gel of the type K_2 (Cu,Ni,Zn) [Fe(CN)$_6$]nSiO$_2$. They were prepared by coprecipitation from the respective Fe(III) cyanides with silica gel in the presence of a reductant. Tl$^+$ is exchangeable versus K$^+$. By oxidation with CaOCl$_2$ or NaOCl, Tl could be desorbed again (Alikin et al., 1984).

5.2.3. Soil Fertilization

The Tl contents of fertilizers and soil additives (e.g., clays, limestone; see Table 4) is generally low, thus enrichment can be practically excluded, if usual application rates are considered (Eschnauer et al., 1984). Analysis of selected environmental matrices led even as early as 1960 to the conclusion that fertiliz-

Table 4   Tl Contents in Fertilizers, μg/g

Compound	Geilmann et al., 1960	Eschnauer et al., 1984	Severin et al., 1990	Sager, unpubl., 1997	Lottermoser and Schomberg, 1993
K salts	0.015–0.31	0.003–0.57	0.05–0.18	<0.2–1.1	—
NaCl	0.008	—	—	—	—
Ammonium nitrate lime	—	—	0.04	<0.1	—
Various NPK	—	0.13–0.27	0.09–1.80	<0.1–0.95	0.2–0.7
Basic slag potash	—	—	0.23	—	0.1–0.4
Manure	—	0.01–0.06	0.07–0.18	—	—
Sewage sludge	—	—	0.09	—	—
Compost	—	—	0.26	—	—

ing with K salts may continously enrich arable soils with thallium (Geilmann et al., 1960).

In Germany, a threshold value for thallium of 1 mg/kg has been imposed on the thallium contents of fertilizers, because it frequently occurs enriched in K salts. In aqua regia extracts of fertilizers, Tl occurrence is usually moderate (Severin et al., 1990).

Amendments of nutrients to the field should be selected in such a way as to ensure minimum load of persistent unwanted substances. With respect to the input of 45 tons/km^2 N, 8 tons/km^2 P, and 16 tons/km^2 K, the annual input of Tl/km^2 was estimated to be 14 g in case of mineral fertilizer, 7–30 g from liquid manure, 20 g from sewage sludge, but 325 g from compost (Severin, et al., 1990). In 1981, sewage sludges from eight different sites in Northwest Germany, representing 10 million inhabitants, contained thallium within 0.10–0.21 μg/g dry weight, with an average of 0.14. With respect to the mean abundance in the earth's crust of about 0.75 μg/g, this means significant depletion (Heinrichs, 1982). Similarly, Severin et al. (1990) found 0.09 μg/g dry weight in sewage sludge, and 0.26 μg/g dry weight in compost. In Tl-contaminated soils, nitrifying bacteria are inhibited, which may affect agricultural production (Qi et al., 1992)

5.3.   EFFECTS ON GREEN PLANTS

5.3.1.   Ambient Levels and Tolerances

Accumulation and resistivity behavior of green plants toward thallium can vary widely and depend on kind and age of the plant as well as the levels of concomitant nutrient and trace elements.

5.3.1.1. Field Studies

Plant uptake of trace metals depends on species-specific properties, as well as on soil humus, clay size, pH, and cation exchange capacity. Usually, the thallium concentration is highest in seedlings, and drops within the life of the plant, except for winter wheat (*Triticum aestivum*), which germinates in November. On a dry weight basis, thallium contents in plants was lower than in soil, except for rape (*Brassica napus*), which exhibited a thallium uptake rate 1000 times more than for others (barley, oats, corn, wheat) (Lehn and Bopp 1987).

For the growth of oats on sand in a greenhouse, thallium toxicity in terms of dry weight in the plant is medium. Retardation of growth took place at more than 20 μg/g dry plant tissue (Davis et al., 1978). For bush beans (*Phaseolus vulgaris*), grown on loamy sand soils pH = 6.8, Tl accumulation in the leaves was not affected by Cd, Cr, Ni, and V (Kaplan et al., 1990). Reduction of crop yield for summer barley occurs at 20 μg/g, for sunflowers, maize, and Salat at 100 μg/g, for spinach at 280 μg/g, and for green cabbage only at 500 μg/g (Makridis and Amberger, 1989b). In bush beans, Tl preferably went into the roots, in green rape it went into the leaves, for peas and bush beans into the stalks. The transport capability seems to be influenced in different ways. Tl administered led to a slower uptake of K (Makridis and Amberger, 1989b).

Addition of highly water-soluble dust from a cement kiln to increase the Tl concentration from 0.4 to 0.9 μg/g within five steps did not significantly change the crop yields for green rape, bush beans, and rye grass. At pH 5.6, the Tl transfer into the plant was about 50% higher-than at pH 6.2. Green rape strongly accumulated Tl, and reached even at an addition of 0.05 μg/g to the soil, higher contents in its dry mass, than the soil itself. Thallium in bush beans and rye grass rose marginally. A linear positive relationship was established between Tl offer and Tl uptake (Makridis and Amberger, 1989a). In Baden Württemberg, Germany, 133 green plants grown near cement plants were in the range 0.02–0.08 mg/kg, mean 0.035 mg/kg, whereas 74 green plants grown at sites of historic mining activities, contained Tl at 0.1–35 μg/g (mean 2.05) (Scholl, 1980).

In mushrooms sampled at various sites of Europe (421 species, 1107 samples), 85% of the samples contained less than 0.25 μg/g Tl, the rest ranging up to 5.5 μg/g. With respect to the Tl contents of the soils they grew in, depletion to about 20% took place, and neither any tendency of differentiation between different classes nor specific distribution patterns between different parts of the mushroom could be detected (Seeger and Gross, 1981). In 33 mushroom samples from Japanese forests, Tl contents ranged 0.001–0.34 mg/kg dry weight, with a mean of 0.07 mg/kg. The Tl concentrations in mushrooms over the respective soil concentration (transfer factor) was 0.2. Unlike for Cs, V, Cd, Pb, and Cu, no species-specific Tl accumulations were observed (Yoshida et al., 1996).

5.3.1.2. Effects in Nutrient Solutions

Nutrient solutions provide a steady state of supply of nutrients and trace elements to the plant roots, avoiding discussions about plant-available fractions in soils. In culture solution, effects on plant growth have been reported for barley at >0.2 μg/mL, for tomato at >0.5 μg/L, for maize at 2.5 μg/L and for rape >10 μg/L (Allus et al., 1987). Barley plants (*Hordeum vulgare*) and rape plants (*Brassica napus*) were transferred to Hoagland's nutrient solution, spiked with $TlNO_3$ or Tl_2SO_4, and harvested after 4 weeks. $TlNO_3$ and Tl_2SO_4 gave the same results. As the upper critical limit of Tl contents within the plants, 86 μg/g in barley shoots, 21 μg/g in barley roots, but as much as 3000 μg/g in rape shoots were found; rape also strongly accumulates thallium from soil (see below). In barley, Tl was higher in roots than in shoots, and in rape it was the reverse (Allus et al., 1987). For *Lolium perennne* (perennial ryegrass) seedlings, germinated in purified sand and grown in Hoagland's nutrient solution, at minimum lethal concentration, thallium was found to be as high as 250 μg/g dry weight in shoots and 2000 μg/g in roots. At below 1 μg/g in shoots, and below 2 μg/g in roots, growth was not effected (Al-Attar et al., 1988). Biomass of barley was significantly reduced when grown at 0.02 mg/L, whereas in rape toxicity symptoms were not noted even at 10 mg/L (Kaplan et al., 1990).

Soybean plants were grown in nutrient solution within a greenhouse under fluorescent light at pH 5.5 and at 0.55 and 1.00 mg/L Tl. Chlorotic areas along the leaf veins darkened roots were noted as early as 8 days after amending the solutions with Tl. After 33 days, the upper leaves were chlorotic and stunted. The Tl addition resulted in significant reductions in all biomass parameters measured for soybean at the 0.55 and the 1.00 μg/g level. Roots were most affected; most of the adsorbed Tl remained in the root zone. Whereas Tl in roots was positively correlated with the Tl application mode, Tl in the leaves remained constant. As a result of Tl treatment, concentration of K in soybeans increased, but Ca and Mg were deficient (Kaplan et al 1990).

In hydroponic culture, K-deficient 15-day-old cucumber seedlings were more sensitive to thallium than those receiving additional K. Above the 200 mg/L level, Tl inhibited plant growth, and at 2 g/L, K deficiency symptoms developed. From Tl distribution between cell organelles it was concluded that processes dependent on cell multiplication may be more sensitive to Tl than those of cell enlargement and differentiation (Siegel and Siegel, 1976).

5.3.2. Accumulating Green Plants

In Tl-contaminated areas, white cabbage, kohlrabi, and green salad highly accumulated Tl, whereas carrots and celeriac were not affected (Scholl, 1980). Rape and cabbage have high affinities for thallium and therefore are not recommended to be grown within a Tl-polluted area. As a tolerance level, 1 mg/kg dry weight for rape and 0.1 mg/kg fresh weight for fruits and vegetables have been establiashed from the German Bundesgesundheitsamt (Crössmann, 1994). At sites around a point source emitting flue dust of high water-soluble contents of thallium, Tl in soils ranged within 0.8–20 μg/g, but the corresponding plants were below 1 μg/g dry weight. Cabbage of any kind, however, and kohlrabi, significantly accumulated thallium. For all kinds of vegetables investigated, leaves contained more than the roots. The amount accumulated in the plants was highly variable between sites and plant species and was not correlated with the amount extractable from the corresponding soil with conc. HNO_3 (Hoffmann et al., 1982). Even for strongly accumulating green plants, depletion of the soil by plant growth is too low to achieve sanitation of heavily polluted sites (Hoffman et al., 1982). The thallium content of the rape plants correlated with Tl soil content, especially with the fraction exchangeable with 1 M ammonium acetate, pH 7, and to a minor extent with the phases reducible with hydroxylamine/HNO_3 at pH 2 (Lehn and Schoer 1987).

5.3.3.   Speciation Inside the Plant Cell

In cytosols of Lemna minor, obtained by extraction of stock cultures in Tris buffer/NaCl, pH 7.0, Tl speciation could be investigated by gel permeation chromatography. More than 80% of the Tl load moved to species of M <1500 g/mol, and a small amount to M = 150,000 (Kwan et al., 1990).

Rape (*B. napus*) has extraordinarily high accumulation and tolerance capabilities versus Tl. Rape leaves were homogenized in Tris buffer, pH 8, at +5°C, and ultracentrifuged to separate a cytosol and a pellet fraction. In winter rape seedlings (January), thallium was quantitatively soluble. In May, in leaves from unpolluted rape the ratio of Tl contents between pellet and cytosol was about 1:2. The cytoplasmic fraction contained 84% of whole Tl. Gel filtration revealed only one Tl species at M = 3800 g/mol, and no free ionic species (Günther and Umland, 1988).

5.4.   EFFECTS ON ANIMAL PRODUCTION

Tl^{3+} is rapidly reduced to Tl^+ in biological systems and thus acts like Tl^+. After intake in rats, distributions between various tissues, down to renal and hepatic subcellular organelles, were the same. The most important organometallic ion is $(CH_3)_2Tl^+$. It is lipophilic and stable under physiological conditions. It can be formed by methylation of Tl with methylcobalamin. $(CH_3)_2Tl^+$ acts like the inorganic ions but is not resorbed well (Sabbioni et al., 1980).

5.4.1.   Ambient Levels and Tolerances

Normally, thallium concentrations encountered in animal tissue are lower than 10 ng/g dry weight. Pig liver (German Environmental Specimen Bank) had 2.96 ± 0.28 ng/g dry weight (Waidmann et al., 1990). In three laboratories, baseline data of thallium levels in human urine, whole blood, and serum have been determined from more than 400 unexposed subjects living in Lombardy, Italy. Subjects suffering from alcoholism, cardiovascular disorders, mental diseases, and cancer, as well as pregnant women and smokers consuming more than 10 cigarettes per day were excluded: 0.42 ± 0.09 $\mu g/L$ were found for urine, 0.39 ± 0.05 $\mu g/L$ for whole blood, and 0.18 ± 0.01 $\mu g/L$ for serum in healthy Italians (Minoia et al., 1990). Nonexposed humans in Germany had Tl urine levels of 0.30 ± 0.14 $\mu g/L$ (Dolgner et al., 1983).

Thallium contents in homogenized freeze-dried muscular tissue of bream from Lake Constance was found to be 5.08–8.61 ng/g, from Lake Belau 0.99–0.84 ng/g, and from the Saar River 0.78–1.92 ng/g (Waidmann et al., 1994).

In mussel tissue from Lake Constance, Tl varied between 75 and 265 ng/g. But in the lakeland district of Bornhoeved, Denmark, which can be regarded as a background region, only 7 ng/g were found (Waidmann et al., 1994).

5.4.2. Exposure and Toxicity in Domestic Animals

Thallium salts were used for rat and mice poisoning for a long time; therefore many experiments about toxicity symptoms, cell physiology, pharmacokinetics, and therapy have been done with rats and mice. As this goes beyond the scope of agriculture, the reader may refer to the recent reviews by Forth (1993) and Sager (1994) [older reviews include Zitko (1975), Friberg et al. (1979), and Formigli et al. (1985)].

For mammals, the 50% lethal dose (LD_{50}) lies between 5 and 70 mg/kg body weight. Intoxications exert main effects in the nerve and digestion systems, accompanied by renal necrosis and loss of hair, as well as disorders in the Na/K metabolism. For guinea pigs, dogs, and men, single doses of 15 mg/kg body weight are totally lethal without therapeutical treatment. Similarly, 50% lethality has been obtained by single doses of 15 mg/kg body weight for rats, and 16–27 mg/kg body weight for mice.

In cats, at doses of 4 mg/kg body weight, hypotony and ataxy occur, which are mainly due to pathological changes of primary sensoric neurons. Further, dogs and cats fed on Tl-containing rodenticides, hemorrhagic gastroenteritis together with hepatic and renal damages emerged. If they survive longer, further symptoms develop like loss of fur, bloody lesions of the skin, tremor and paralysis of muscles. In dogs, Tl-induced testicular atrophy has been described, with simultaneous premature release of germinal cells into the seminiferous tubule. For rabbits, the minimum lethal dose is 20–30 mg/kg body weight. A daily dose of 0.25 mg/kg body weight, given to domestic rabbits over a period of 6 months, caused behavioral disorders, aggressivity, crippled legs, diarrhea, hair loss, as well as histologic alterations of the liver, kidney, and stomach membranes (Török and Schmall, 1982).

In order to investigate the action of thallium-contaminated food on domestic animals, 15 sheep of 25 ± 2 kg body weight were supplied with Tl$^+$ via potable water, up to 1.0 mg Tl/kg body weight a day. Within 15 weeks, even a daily intake of 0.03 mg Tl/kg body weight, corresponding to 1.5 μg/g in the food dry weight, led to significant accumulation in all tissues studied and to rather homogenous distribution among the tissues.

Similarly, eight bulls at a farm within 800 m distance to a Tl emission source, and fed with locally grown food, were investigated. The calculated daily intake from the contaminated local food was 10 mg, mainly from maize and hay, whereas the load from barley and soya was much lower. The source of Tl ingestion in this case was flue dust, partially water soluble. During a 6-month feeding period, contrary to the experiment with sheep, Tl in the bulls got significantly accumulated in the kidneys. These results indicate that the Tl contents of animal food should not exceed 1 mg/kg dry weight in order to avoid accumulations of Tl in edible tissues (Hapke et al., 1980).

For mammals and men, urine and hair samples are easily accessible indicators to monitor thallium exposure, because excess uptake is preferably excreted this way. First data for Tl contents of presumably noncontaminated human hair and finger nails were given at 7–51 ng/g and 40–74 ng/g, respectively (Geilmann et al., 1960). Hair of various mammals was found within the same range (3–60 ng/g). Thallium contents of scalp hair from cattle in Baden Württemberg, Germany, ranged from 0.08 to 2.7 μg/g as supplied, plus one outlier at 6.8 μg/g. Thallium in scalp hair from calves was higher than from old animals (Scholl, 1984).

For chicken embryos, thallium is highly teratogenic, producing achondroplasia, leg-bone curvature, microcephaly, and reduced fetal size (Formigli et al., 1985).

Thallium-containing effluents may harm fishing and fish farming. In natural waters, Tl is not precipitated as the carbonate nor as the hydroxide and is also hardly bound to soluble humic material. Therefore, toxicity is largely independent from water characteristics, such as water hardness, dissolved organic carbon, or suspended matter (Nehring, 1962; Zitko et al., 1975).

Zooplankton, such as daphnia, react more sensitively than the fish that feed on them. Symptoms of thallium toxicity are sudden motions of legs as well as paralysis. Thus, Tl emissions can harm fishery, without toxicity symptoms on the fish themselves (Nehring, 1962). Freshwater fish exert toxicity symptoms after 1–2 days. This means sudden moves of fins, short breath, disturbed balance, and turning aside. Further on, increased blood pressure can lead to loss of blood via gills and skin. Poisoned fish usually recreate in clean water. Short-term amounts of 2 mg/L Tl do not harm fish; chronic intoxications, however, occur at much lower concentrations (Nehring, 1962; Zitko et al., 1975). For the Atlantic salmon, chronic intoxication starts at 30 μg/L; increase in thallium concentration is only moderately more lethal, contrary to Cu and Zn. Significant accumulation of thallium occurs in muscle, liver, and gills (Zitko et al., 1975). For fathead minnows (*Pimephales promelas*) at 25°C, pH 6.7–7.2, and good oxygen supply, the LC$_{50}$ (50% lethal concentration) for adults was 860 μg/L. For embryos, 720 μg/L was lethal, but 200 μg/L had no effect. Survival of larvae, however, was significantly reduced at 40 μg/L (Le Blanc and Dean 1984).

5.4.3. Thallium in Agricultural Products

Milk from Baden Württemberg contained Tl within the range 0.010–0.033 μg/g dry weight (Scholl, 1980). Red wines from Ingelheim, Rhineland, Germany, contained more Tl (average 0.21 μg/L) than the white wines (0.12 μg/L) grown within the same area (Table 5). With respect to dry weight, grapes and peels had the largest, and cernels the lowest Tl concentration, whereas Tl in stalks from red-wine grapes was higher than from white-wine grapes (Eschnauer et al., 1984). As tobacco contains Tl above the average (24–100 ng/g), smokers suffer additional Tl exposure (Geilmann et al., 1960).

Dietary habits were found to be related to the increased intake of Tl. Consumption of home-grown vegetables and fruits grown in the vicinity of a cement plant, which emitted dust of high Tl contents, led to increased Tl uptake. This was monitored by 24-h urine samples. Grass and soil monitoring provided evidence of substantial fallout of dust containing Tl in the cement plant area. Major route of increased intake for humans was consumption of locally grown food. The pulmonary route of uptake was insignificant. Even children attending a kindergarten at about 0.5 km distance from the source had no higher exposure than adults (Dolgner et al., 1983).

5.5. TRENDS IN ANALYTICAL TECHNIQUES FOR DETERMINATION OF THALLIUM

In industrialized countries, rising costs of manpower and improvements of instrumental analytical equipment lead to increasing automatization, and the trend toward multi-element techniques. In parallel, the selectivity of determinations is increased in order to avoid separation procedures. Increasing selectivity and sensitivity of spectroscopic methods may render separation and enrichment techniques unnecessary in many cases.

Table 5   Thallium in Noncontaminated Agricultural Products

Product	Range, ng/g	Literature
Plants for animal feeding	0.02–0.08	Scholl, 1980
Vegetables (edible parts)	0.03–0.20	Scholl, 1980
Red wines	0.09–0.48	Eschnauer et al., 1984
White wines	0.04–0.3	Eschnauer et al., 1984
Milk	0.010–0.033	Scholl, 1980
Bream (freshwater fish)	0.8–8.6	Waidmann et al., 1994

Table 6   Currently Available Standard Reference Materials (status N: not certified)

Material	Standard	Status	Conc., μg/g
Blend coal	BCR-CRM 040	N	0.79
Gas coal	BCR-CRM 180	N	2.2
Bovine serum	NIST-SRM 1598	N	0.0004
Skim milk powder	BCR-CRM 150	N	0.001
Skim milk powder	BCR-CRM 151	N	0.0008
Dogfish muscle	NRCC-DORM-2	N	0.004
Rye bread flour	CSRM 12-2-05	N	0.02
Wheat bread flour	CSRM 12-2-04	N	0.05
Pine needles	NIST-SRM 1575	N	0.05
Lucerne P-Alfalva	CSRM 12-2-03	N	1
Eutric cambisols soil	CSRM 12-1-07	N	0.2
Orthic luvisoils soil	CSRM 12-1-08	N	0.2
Rendzina soil	CSRM 12-1-09	N	0.2
Yellow-brown soil	GBW 07403	N	0.48
Soil	GBW 07409	C	0.58
Loess	GBW 07408	C	0.59
Chestnut soil	GBW 07402	C	0.62
Soil	GBW 07410	C	0.62
San Joaquin soil	NIST-SRM 2709	C	0.74
Limy yellow soil	GBW 07404	C	0.94
Podzolitic soil	GBW 07401	C	1.0
Montana I soil	NIST-SRM 2710	N	1.3
Yellow-red soil	GBW 07405	C	1.6
Soil	GBW 07411	N	1.7
Yellow-red soil	GBW 07406	C	2.4
Montana II soil	NIST-SRM 2711	C	2.47
Dry soil	EPA-CRM 003-50	N	20

To check the validity of the applied method, a selection of Standard Reference Materials is compiled in Table 6. Note that there is hardly any material available for analysis of fruits, green plants, or fertilizers. Thus, the analyst largely has to prove his method by himself, e.g., to take care for regain as well as correction for background and blanks.

Restrictions for disposal of hazardous waste restrict solvent extraction methods and nuclear methods. Within this context, it is not possible to cover all techniques utilizable in environmental analytical chemistry. Therefore, only recent developments and trends of methods often used in routine laboratories, which make a great deal of environmental data, are outlined.

For a survey of analytical methods, graphite furnace atomic absorption spectrometer (AAS), plasma emission spectroscopy, and mass spectrometrical and electrochemical techniques will be discussed in detail. They are probably the most frequently used analytical techniques used in modern routine laboratories. Unlike the situation about 15 years ago, graphite furnace AAS and inductively coupled plasma mass spectrometer (ICP-MS) have made such progress suitable for trace thallium determination in the routine laboratory.

Proton-induced x-ray emission (PIXE) also offers the possibilities of determination of ambient levels of thallium within multi-element screening, but its use is limited to a few places of the world, where proton accelerators are installed. Spectrophotometric determinations require separation of most of the inorganic matrix, as well as complete destruction of organic material. X-ray fluorescence is too insensitive for analysis at the level of average crustal abundance. Laser ablation atomic fluorescence seems to be the most sensitive technique available for thallium, but it is not in routine use now.

Separation methods, which enable speciation studies, still require much research. Chromatographic techniques with element-selective detection may offer an additional dimension of knowledge in the future.

5.5.1. Electrothermal Atomization–Atomic Absorption Methods

In electrothermal atomization techniques, heating at a reducing surface (usually graphite) should transform the analyte to the elemental state, with subsequent formation of monoatomic vapor. Thallium has a tendency to form thermally stable compounds more rapidly than the reduction occurs, which leads to significant decrease of the analytical signal. These are $TlCl$, TlF, $TlBr$, Tl_2O, Tl_2O_3, and even $TlNO_3$.

Though pure $TlClO_4$ is thermally stable to 500°C, use of digests from perchloric acid yields low sensitivity because it decomposes to the thermally stable compounds Tl_2O_3 + $TlCl$ + Cl_2 + O_2 (Solymosi and Bánságy, 1971). The interference by chloride is due to the formation of $TlCl$ in the gas phase during pyrolysis or atomization stage of the furnace heating cycle, and its loss by diffusion prior to its complete dissociation. Use of a dual-cavity platform, to which the analyte was added on one side, and the chloride on the other, proved formation of $TlCl$ in the gas phase as well as on the tube surface (Qiao et al., 1993; Welz et al., 1988). Formation of thermally stable molecules limits the maximum charring temperature to about 400°C for samples in dilute HCl, and to about 500°C for samples in dilute HNO_3 or H_2SO_4 solution. In order to determine Tl in digests or extracts from real samples, either separations or modifications of the analytical conditions have to be performed. If only D_2 background compensation is available, smoke from Fe, P, and alkaline earths will deteriorate the signal in real sample solutions. Thus, addition of a matrix modifier should be avoided, which causes additional background, and standard addition has to be applied. In order to improve detection limits, separation and enrichment steps have been made, to approach the sensitivity of the pure standard solution (reviewed by Sager, 1986, 1992a).

After gel chromatographic separations of various Tl compounds in Tris buffer, pH 8, Tl could be directly determined versus a calibration graph (Günther and Umland 1988).

The Tl line at 276.8 nm exhibits an anomalous Zeeman pattern. The relative absorbance passes a maximum at 9 kG field strength, and declines at higher field strengths. Background correction utilizing the longitudinal Zeeman effect thus significantly improves the analytical signal in presence of matrix P and Fe, with respect to the transversal Zeeman mode.

Thallium does not form carbides during the determination cycle.

5.5.1.1. Suitable Decomposition Procedures

Digests obtained with aqua regia or $HClO_4$, and also any solutions of the sample in HCl, are unsuitable for direct measurement in graphite furnace AAS because of significant depressions of the thallium signal. Incomplete background correction may lead to erroneously high results (author's experience). Destruction of organic material in the muffle furnace leads to nonpredictable losses because, unlike Zn, Cd, or As, the Tl is volatile as Tl_2O from alkaline ashes also. Recommended methods include HNO_3 leach (e.g., soils, inorganic salts), pressure decomposition with HNO_3 (e.g., coal), or combustion methods (author's experience), unless separation steps are applied.

For routine analysis of agricultural soils, samples were leached with hot HNO_3 (to expell the halogenides), and the filtrate submitted to graphite furnace AAS without further matrix modification. Uncoated tubes with inserted platforms and Zeeman background compensation were used (Hofer et al., 1990).

5.5.1.2. Separations

Thallium in digests of sediments, soils, and coals, as well as in leachates from HCl or NH_4Cl, can be easily separated from the matrix by solvent extraction. Prior to introduction into the furnace, it is advantageous to strip the organic layer and to acidify the resultant aqueuous solution with HNO_3 or H_2SO_4 (Günther and Umland, 1988; Sager, 1992b). After coprecipitation with MnO_2, for example, the sample should be dissolved with HNO_3.

After volatilization and preconcentration from rock samples within a stream of 20% H_2 in N_2, Tl was dissolved from the condenser with $1 + 10$ HNO_3 and determined without the need of further matrix modification (Heinrichs, 1979). Other coprecipitation, solvent extraction, and volatilization methods, as well as ion exchange and electrolysis, have been already intensely reviewed (Sager, 1986, 1992), but the frequency of their application clearly declines.

5.5.1.3. Matrix Modifiers

Within the last years, a clear trend to avoid laborious separation steps and to improve background compensation can be observed. Matrix modifiers are added to retain Tl in the charring step at temperatures high enough to volatilize matrix, such as sulfate and phosphate from the sample, or vice versa, to retain the matrix till the end of the Tl signal.

For samples low in alkali, alkaline earths, Fe, and P, replacement of halogenids during the charring step by sulfuric acid or nitric acid can be sufficient, even with D_2 compensation, for example, after separation of Tl via a halogenide. If matrix modifiers such as H_2SO_4 or $Pd(NO_3)_2$/ascorbic acid mixture are used, background compensation utilizing the Zeeman effect showed distinct advantages over compensation with D_2. National Bureau of Standards (NBS) pine needles were taken as a sample for demonstration (Wegener et al., 1992).

Addition of Pd^{2+} as a matrix modifier allows one to raise the pyrolysis temperature up to 1000°C. Palladium stabilizes the thallium by delaying its absorbance signal (Liang and Ni, 1994). Much more effective stabilization by Pd is seen when Pd has been prepyrolyzed at 900°C before adding the sample. The breakdown of Pd–Tl within the intermediate alloy Pd_2Tl is the rate–determining step. Palladium itself vaporizes at higher temperatures than the analyte (Liang and Ni, 1994). Addition of ascorbic acid to the Pd modifier improved absorbance signal slightly and reduced the Cl interference. Some urine samples showed a depressed signal, which could not be overcome, however (Collett and Jones, 1991).

Palladium and Pd/Mg nitrate modifier did not significantly diminish the signal depression of the analytical signal by <0.01% perchloric acid; for 3% $HClO_4$, the signal was only 50% in spite of presence of the modifier (Welz et al., 1988). Using NH_4–EDTA + NH_3 + $Pd(NO_3)_2$ as a matrix modifier completely suppressed the interference of $HClO_4$, in pyrocoated tubes, even with D_2 compensation. Solid samples were decomposed with HNO_3/$HClO_4$/HF, finally dissolved in 0.02M-HNO_3, prior to addition of the modifier (Zheng and Wang 1995).

Addition of Li salts as well as of H_2SO_4 reduced interferences from chloride (Welz et al., 1988). As the dissociation energy of Li–Cl is higher than for H–Cl, Na–Cl, and Mg–Cl, Li is capable to mask chloride in the atomization cycle. Sodium chloride as such is stable up to 800°C (Qiao et al., 1993). In [1% $LiBO_2$ + 1% H_2SO_4 + 5% HNO_3], no charring losses of thallium platforms within pyrocoated tubes were observed. Thus, borate melts, dissolved in 10% HNO_3, could be submitted to graphite furnace AAS (with Zeeman correction) (Grognard and Piolon, 1985).

Direct determination of Tl in a HNO_3 solution of fertilizer phosphates (up to 27% P_2O_5) in a transversally heated graphite furnace was possible after addition of 5% aqueous ammonium molybdate. The analyte was stabilized up to 750°C, and the modifier sufficiently retarded the smoke from the phosphate till the end of the Tl signal (Sager, unpublished).

5.5.1.4. Other Modifications

A coating of TaC significantly lowered the matrix effect observable in pure HCl solutions, whereas it had only marginable influence on $HClO_4$. Coating with TaC was achieved by soaking with a Ta solution containing fluoride and oxalate, with subsequent heating to 1000°C. Carbidization with Zr in HCl solution was not satisfactory. After carbidization with W in 0.5% NaOH, influence of $HClO_4$ was significantly lowered (Hamid et al., 1991). Chloride interference could also be overcome by use of 5% H_2 in Ar as purging gas (Welz, et al., 1988).

5.5.2. Atomic Emission Spectroscopy from Inductively Coupled Plasmas

The thallium emission spectrum is entirely due to the neutral atoms, with all transitions terminating in the 6p ground state. The most sensitive emission line in the inductively coupled plasma (ICP) is probably at 190.864 nm, but vacuum conditions are needed. For spectrometers operating in air, four emission lines are useful: 276.787, 351.924, 377.572, and 535.046 nm (Boumans, 1989; Dorado Lopez et al. 1982). The sensitivity, however, is rather insufficient for any of the analytical lines used to get reasonable results for noncontaminated matrices in multi-element screening. Thus, the direct application to decomposition solutions of rocks, soils, and biological samples is limited. At an excitation frequency of 50 MHz, detection limits are better than at 100 MHz (Boumans, 1989). As thallium is easily excitable, optimum signals to background ratios are obtained at low powers.

At 276.787 nm, spectral interferences may occur from more than 100-fold excess of Fe (276.75 nm), as well as a second-order interference with Ba (553.548 nm) (Anderson and Parsons, 1984). At 377.572 nm, which is often used in direct current arc excitation methods, may not be resolved from Ca (377.59 nm), N (377.56 nm), Ti (377.61 nm), and V (377.62, 377.57, and 377.52 nm) (Botto, 1981), and above all Ar at 377.545 nm (Anderson and Parsons 1984). Therefore, no practical application of this line has been found in ICP-atomic emmision spectroscopy (AES). At 535.046 nm there are hardly spectral interferences. Within the last years, to the author's knowledge, environmental data at noncontaminated levels obtained by this technique did not appear in the scientific literature.

5.5.3. ICP with Mass Spectrometric Detection (ICP-MS)

Especially for heavy elements, ICP-MS seems to be a promising technique, which has developed rapidly within the last decade. Digests of real samples have to be prepared in dilute HNO_3. Degradation of $HClO_4$ yields many metastable species, which may interfere with the ions to be detected in the mass spectrum. Easily ionizable elements lower the signal because they shift the ion/atom ratio, but Tl is less affected than Y and Li (Gregoire, 1987). In 0.7% HNO_3 solutions, a detection limit of about 1 $\mu g/L$ was reached in multi-element runs and applied to detect blanks from blood sampling and storage tubes (Paudyn et al., 1989). In the routine analysis of lake waters, a detection limit of 0.1 $\mu g/L$ has been achieved, which is due to the instrument and not to the blanks.

Rock samples were digested with $HF/HClO_4/HNO_3$ in a microwave furnace, fumed in the open with HNO_3, and finally dissolved in 1:10 HNO_3, and diluted 1 + 4 before measurement. Samples of 20 $\mu g/L$ Rh and/or In were used as internal standards. As limit of determination, 0.5 μg/g for Tl was achieved for 10-s integration time (Yoshida et al., 1996). For determination of thallium in plasma and urine, the samples were just 1 + 4 diluted with 0.144 HNO_3, and fed to a pneumatically operating Meinhard nebulizer, for 40-s of integration time. The standards were matched with Na/K/Ca/Mg. Calibration curves were linear, and Eu was taken as internal standard. In urine samples from healthy subjects of the Paris region, Tl and Bi could not be detected (Mauras et al., 1993).

Sample introduction into the ICP by electrothermal vaporization leads to a dramatic improvement in analyte transport efficiency. The samples are volatilized from a conventional graphite furnace and transported as a dry aerosol to the torch of the ICP-MS. Thus, the limits of detection are improved 10–50-fold over conventional solution nebulization. During the dry and char stages, the Ar has to remove interferent vapors through the dosing hole. As all molecules (TlCl, e.g.) are cracked in the ICP afterward, no modifier is needed. In arctic snow samples, 0.57 pg/g could be detected, which was 345 times more than with AAS measurement from the same furnace (Sturgeon et al., 1993).

Similarly, for determination of Tl in ambient dust samples, dust was sampled by graphite filters or graphite impaction plates, electrothermally atomized within a graphite furnace, and subsequently detected by ICP-MS coupled on line. About 1 pg/m^3 could be detected (Tilch et al., 1996).

5.5.4.   Electrochemical Techniques

Electrochemical techniques have been the most sensitive determination methods for thallium for a long time, achievable with rather cheap equipment, but skillful operators and elaborate procedures are needed to avoid systematic errors. As the analytic signal depends on the actual diffusion of the analyte within the sample solution, standard addition techniques are essential. Sufficient sensitivity is gained by electrodeposition at the working electrode, with subsequent redissolution. For thallium, the reversible redox couple Tl$^+$/Tl$^\circ$ at about -0.5 V versus saturated calomel electrode is used (Bard, 1975).

Thallium forms an amalgam with Hg, which is stable within a broad concentration range, and Tl$^+$ is deposited about 1.6 times faster than Tl^{3+}. To achieve correct standard addition conditions, the analyte must be in the monovalent state, like the standard solution added. Reduction of Tl to yield monovalent Tl in any case can be done with hydroxylamine or sulfite. In the presence of nitrite, the NH_2OH causes the formation of gas bubbles, which interfere, when they move to the electrode surface. Infinite tolerance versus alkali, alkaline earths, and halogenides work well for the analysis of seawater, brines, and biological materials. For thallium, the multi-element capabilities of the method cannot be used because other electroactive species have to be masked completely to achieve sufficient selectivity. In aqueous solutions, the half-wave potential of Tl is close to Pb, Cd, In, and Sn, depending to some extent on the basic electrolyte. As thallium complexes are rather weak, the electroanalytical thallium signal is nearly nonaffected by complexants in solution. Masking of these ions has been known for a couple of years to be effective with complexants such as EDTA, or surfactants, which shift or diminish all other peaks except Tl.

With pH increasing, complexing capacity of any basic electrolyte increases, but the signal obtainable from the Tl also decreases; thus compromise conditions have to be found. As the literature up to 1991 has already been intensely reviewed (Sager, 1986, 1992a), this text is focused on general practical experiences and recent trends.

Generally, electrochemical techniques have been largely applied to natural waters, as well as to acid digests of alloys. In these cases, possible electroactive species are known, and they are successfully masked or separated. Numerous works have appeared, dealing with electrochemical determination of Tl in matrix Bi, Cd, Cu, In, Ni, Sn, and even Pb (which yields the same position of stripping peak in dilute mineral acid solution). EDTA, citrate, or tartrate as conducting electrolytes are well suited to complex interferents. Large amounts of Cu can be masked with polyethylenglycol, Fe(III) with hydroxyl amine (Gemmer-Colos et al., 1981), ascorbic acid, or oxalate/Rokaphenol N-3 (Lukaszewski et al., 1987).

If in the stripping, the potential of a reversible redox couple present in the solution is passed, background current is much enhanced (Kryger, 1980; Liem et al., 1984). On the whole, there are four strategies to avoid those hardly reproducible cross interferences: separation at high temperatures (combustion, volatilization as oxide or chloride), high pressure ashing, separation from the matrix, or change of the sample solution in order to perform stripping of deposited metals into pure electrolyte.

The direct use of solutions from the decomposition of rock samples, however, is hampered by larger excess of Cu, Fe, and Ti, which make the Tl peak disappear in the background current (Liem et al., 1984). There were no problems, however, after volatilization of Tl as the chloride and dissolution in dilute H_2SO_4.

Environmental materials, such as soils, plant, and animal tissues, organic fertilizers, sewage sludges, require digestion of the organic materials present. Aqua regia digests and even $HClO_4/HNO_3$ digests are not sufficiently complete to perform deposition and stripping in the same sample solution (author's experience).

The influence of surfactants on the Tl signal is generally very low (see, e.g., Ciszewski, 1985; Lukaszewski et al., 1980), but residual organics from incomplete digestion can yield ghost peaks (author's experience). For coal and rubber samples, an example is given by Gorbauch et al. (1984).

Even inorganic fertilizers (NPK type) may contain some organic substances to form stable grains. After mere dissolution in the basic electrolyte, ghost peaks appeared in the stripping diagram; the fertilizers had to be dissolved with hot HNO_3, and filtered.

5.5.4.1.   Mercury Film Electrode

By electrolytic deposition of Hg at pH <4 on an inert electrode, a layer of metallic mercury is generated, serving as new electrode surface for subsequent use. Within the last years, the mercury film electrode (MFE) has been largely favored over the hanging mercury drop electrode (HMDE) to avoid a pool of liquid mercury, which may be a contamination problem, if Hg has to be determined within the same laboratory, as well as a problem of generating hazardous waste. Recently, electroplating of $HgCl_2$ in 1.3 M HCl on glassy carbon (Cleven and Fokkert 1994; Laar et al., 1994), or from 0.1 M KNO_3 upon epoxy–resin impregnated graphite (Lukaszewski and Zembrzuski, 1992; Lukaszewski, et al., 1996) have been done. The film can be repeatedly used. The potential peak of Tl on an MFE is approximately 150 mV more negative than on an HMDE.

Electrolytic deposition of thallium from soil digests obtained with HF/HCl/HNO_3/H_2O_2, and finally taken into HCl, was exerted within a supporting electrolyte of ascorbic acid/EDTA, pH 4.5. Subsequent stripping was improved by replacement of the sample solution with deaerated 0.05 M EDTA, pH 4.5, within a flow system. Change of solution, into which electrolysis is carried out, avoids voltammetric peaks from nondeposited species and allows the separate optimization of both procedures. A three-electrode flow-through cell enabled electrolytic enrichment from larger sample volumes (100 mL), which circulate through the cell, and also to exchange the electrolyte prior to anodic stripping (Lukaszewski et al., 1996). In the pure stripping electrolyte, the peaks of codeposited Sb, Bi-and Cu appeared well resolved from Tl. A detection limit of 10 ng/g soil was achieved (Lukaszewski and Zembrzuski 1992), as well as 2 ng/L after 90 min preconcentration from 0.05 M EDTA, pH 4.4, electrolyte. A 1000-fold excess of Pb could be tolerated (Lukaszewski et al., 1996).

Among 40 different electrolytes investigated, best separation of Sn, Cd, Pb, and Tl peaks was achieved by stripping in 5 M NH_3/0.5 M KOH, where deposition is difficult (Schulze et al., 1985).

Without a flow system, a detection limit of 0.4 μg/L could be achieved for the determination of Tl in river water samples in the potentiometric stripping mode by means of deposition from 0.05 M EDTA, without further sample preparation. Also, N_2 purging led to a five-fold increase of peak area (Cleven and Fokkert, 1994).

In soil extracts, obtained with 0.24 M HNO_3, Tl was determined on an MFE in a supporting electrolyte of 0.13 M EDTA/0.58 M ascorbic acid, adjusted to pH 4.5. The addition of ascorbic acid removed interferences from Fe. Prior to electrolysis, the final solution was purged with N_2 for 10 min (Laar et al., 1994).

5.5.4.2.   Hanging Mercury Drop Electrode

In environmental water samples, thallium determination in acetate buffer, pH 4.5, in the presence of EDTA has sufficient tolerance versus accompanying

Pb, Cd, and others (Bonelli et al., 1980; You and Neeb, 1983), but detection limits are hardly low enough to reach noncontaminated levels. The methods have been intensely reviewed (Sager, 1986, 1994). As an alternative to the voltammetric stripping determination of thallium in the presence of various complexants, to mask other electroactive cations, adsorptive stripping voltammetric determination of thallium in the presence of o-cresolphtalein has been presented (Wang and Lu, 1993). The complex is preconcentrated at pH 12, -300 mV, and subsequently scanned to -900 mV. After 100-s preadsorption time, surface saturation was reached. The peak increased rapidly upon increasing the pH between 8 and 11; then it started to level off. The differential pulse waveform offered improved signal to background characteristics. Measurements below 5 μg/L required removal of O_2. A five times improvement of detection limit with respect to acetate buffer was achieved. There was good resolution from four-fold excess of Pb, and no signal from Sn and In and further 11 ions. Dissolved organics, like gelatin, significantly decrease the peak. Thus, samples of natural waters had to be treated with ultraviolet light, or addition of fumed silica, to lower the dissolved organic carbon contents (Wang and Lu, 1993).

REFERENCES

Al-Attar, A. F., Martin, M. H., and Nickless, G. (1988). Uptake and toxicity of Cd, Hg and Tl to Lolium perenne seedlings, *Chemosphere,* **17,** 1219–1225.

Alikin, V. V., Volkhin, V. V., and Kolesova, S. A. (1984). Sorption of thallium (1) by composition sorbents, included ferrocyanides of the transition metals and silica gel (russ.). *Izv. A. N. Turkm. SSR, Ser. Fiz. Tekhn. Khim. Geol. Nauk,* 59–63.

Allus, M. A., Martin, M. H., and Nickless, G. (1987). Comparative toxicity of thallium to two plant species, comparative toxicity of thallium to two plant species, *Chemosphere* **16,** 929–932.

Andersson, T. A., and Parsons, M. L. (1984). ICP emission spectra III: The spectra of the group IIIA elements and spectral interferences due to group IIA and IIIA elements. *Appl. Spectrosc.,* **38,** 625–634.

Bard, A. J. (1975). *Encyclopedia of Electrochemistry of the Elements,* Vol. 4. Marcel Dekker, New York.

Batley, G. E., and Florence, T. M. (1975). Determination of thallium in natural waters by anodic stripping voltammetry. *Electroanal. Chem. Interfac. Electrochem.,* **61,** 205–211.

Berthelot, Ch., Carraro, G., and Verdingh V. (1980). Non-destructive multi-element photon activation analysis of river sediments. *J. Radioanal. Chem.* **60,** 443–451.

Bonelli, J. E., Taylor, H. E., and Skogerboe, R. K. (1980). A direct differential pulse anodic stripping voltammetric method for the determination of thallium in natural waters. *Anal. Chim. Acta,* **118,** 243–256.

Botto, R. I. (1981). A proposed dual polychromator system for reducing spectral interferences in multielement ICP, *ICP Inf. Newslett.,* **6**(10), 521–543.

Boumans, P. W. J. M. (1989). Detection limits for lines widely differing in hardness and intensity ratios of ionic to atomic lines measured in 50 and 100 Mhz inductively coupled plasmas. *Spectrochim. Acta,* **44B,** 1285–1296.

Bruland, K. (1983). Trace elements in sea water. In *Chemical Oceanography*, Riley, J. P., and Chester, R. (eds., Academic, Toronto,) pp. 157–220.

Brumsack, H. J. (1977). Potential metal pollution in grass and soil samples around brickworks. *Environm. Geol.*, **2**, 33–41.

Butler, J. S. (1962). Thallium in some igneous rocks. *Geokhim.* (Russ.), **6**, 514–523.

Cheam, V., Lechner, J., Desrosiers, R., Sekerka, I., Lawson, G., and Mudroch, A. (1995). Dissolved and total thallium in Great Lakes waters. *J. Great Lakes Res.*, **21**, 384–394.

Ciszewski, A. (1985). Determination of thallium in bismuth by differential pulse anodic-stripping voltammetry without preliminary separation. *Talanta*, **32**, 1051–1054.

Ciszewski, A., and Lukaszewski, Z. (1985). Electrochemical masking of large amounts of copper in DPASV and the determination of thallium in the presence of a large excess of copper. *Talanta*, **32**, 1101–1104.

Ciszewski, A., and Lukaszewski, Z. (1983). Determination of thallium in lead salts by differential pulse anodic stripping voltammetry. *Talanta*, **30**, 873–875.

Cleven, R., and Fokkert, L. (1994). Potentiometric stripping analysis of thallium in natural waters. *Anal. Chim. Acta*, **289**, 215–221.

Collett, D. L. N., and Jones, S. M. (1991). Determination of thallium in urine by graphite furnace atomic absorption spectrometry. *Atom. Spectrosc.*, **12**, 69–73.

Crössmann, C. (1994). *Transfer und Akkumulation von Thallium bei Gemüse und Obstarten.* *VdLUFA* Kongreβ, Jena.

Davis, R. D., Beckett, P. H. T., and Wollan, E. (1978). Critical levels of 20 potentially toxic elements in young spring barley. *Plant Soil*, **49**, 395–408.

Dolgner, R., Brockhaus, A., Ewers, U., Wiegand, H., Majewski, F., and Soddermann, H. (1983). Repeated surveillance of exposure to thallium in a population living in the vicinity of a cement plant emitting dust containing thallium. *Int. Arch. Occup. Env. Hlth.*, **52**, 79–94.

Dorado Lopez, M. T., Gomez Coedo, A., and Gallego Andeu, R. (1982). Aplicación de la espectrometria de plasma (ICP) al analisis de plomos puros. *Rev. Metal. CENIM*, **18**, 69–81.

Eschnauer, H., Gemmer-Colos, V., and Neeb, R. (1984). Thallium in Wein—Spurenelement-Vinogramm des Thalliums. *Z. Lebensm. Unters. Forsch*, **178**, 453–460.

Formigli, L., Gregotti, C., Di Nucci, A., Edel, J., Sabbioni, E., and Manzo, L. (1985). Reproductive toxicity of thallium. Experimental and clinical data. *Proc. Int. Conf. Heavy Met. Environm. Athens*, 15–18.

Forth, W. (1993). Thallium, cesium and potassium. In *Metal-Metal Interactions*, B. Elsenhans, et al. (eds.). Bertelsmann Gütersloh, Germany, pp. 149–164.

Friberg, L., Nordberg, G. F., and Vouk, V. B. (1979). *Handbook on the Toxicology of Metals*, Elsevier, Amsterdam.

Geilmann, W., Beyermann, K., Neeb, K. H., and Neeb, R. (1960). Thallium, ein regelmäßig vorhandenes Spurenelement im tierischen und pflanzlichen Organismus. *Biochem. Ztschr.*, **333**, 62–76.

Gemmer-Colos, V., Kiehnast, I., Trenner, J., and Neeb, R. (1981). Invers—voltammetrische Bestimmung des Thalliums. *Fres. Z. Anal. Chem.*, **306**, 144–149.

Gorbauch, H., Rump, H. H., Alter, G., and Schmitt-Henco, C. H. (1984). Untersuchung von Thallium in Rohstoff- und Umweltproben. *Fres. Z. Anal. Chem.*, **317**, 236–240.

Gregoire, D. C. (1987). The effect of easily ionizable concomitant elements on non-spectroscopic interferences in inductively coupled plasma-mass spectrometry. *Spectrochim. Acta*, **42B**, 895–907.

Grognard, M., and Piolon, M. (1985). Acid pretreatment fusion method for determination of thallium in city waste incineration fly ash by Zeeman atomic absorption spectrometry. *Atom. Spectrosc.*, **6**, 142–143.

84 Thallium in Agricultural Practice

Gunther, K., Henze, W., and Umland, F. (1987). Mobilisationsverhalten von Thallium und Cadmium in einem Flußsediment. *Fres. Z. Anal. Chem.*, **327**, 301–303.

Gunther, K., and Umland, F. (1988). Speziesanalytik von Cadmium und Thallium in nativen Rapspflanzen. *Fres. Z. Anal. Chem.*, **331**, 302–309.

Hamid, H. A., Al Joboury, M. I., and Mohammed, A. K. (1991). Determination of thallium by furnace atomic absorption spectrometry. *Anal. Chim. Acta*, **243**, 239–245.

Hapke, H. J., Barke, E., and Spikermann, A. (1980). Ansammlung von Thallium in verzehrbaren Geweben von Hammeln und Bullen in Abhängigkeit von der Thalliummenge im Futter. *Dtsch. tierärztl. Wschr.*, **87**, 376–378.

Heinrichs, H. (1975). Die Untersuchung von Gesteinen und Gewässern auf Cd, Sb, Hg, T, Pb und Bi mit der flammenlosen Atom-Absorptions-Spektralphotometrie. Ph.D. Thesis, Univ. Göttingen.

Heinrichs, H. (1977). Emissions of 22 elements from brown-coal combustion. *Naturwiss.*, **64**, 479–481.

Heinrichs, H., and Mayer, R. (1977). Distribution and cycling of major and trace elements in two central European forest ecosystems. *J. Environ. Qual.*, **6**, 402–407.

Heinrichs, H. (1979). Determination of bismuth, cadmium and thallium in 33 international standard reference rocks by fractional distillation combined with flameless atomic absorption spectrometry. *Fres. Z. Anal. Chem.*, **294**, 345–351.

Heinrichs, H. (1982). Metalle in kommunalen Klärschlämmen. *Naturwiss*, **69**, 88–89.

Hofer, G. F., Aichberger, K., and Hochmair, U. S. (1990). Thalliumgehalte landwirtschaftlich genutzter Böden Oberösterreichs. *Die Bodenkultur*, **41**, 187–193.

Hoffmann, G. G., Schweiger, P., and Scholl, W. (1982). Aufnahme von Thallium durch landwirtschaftliche und gärtnerische Nutzpflanzen. *Landw. Forsch.*, **35**, 45–54.

Kaplan, D. I., Adriano, D. C., and Sajwan, K. S. (1990). Thallium toxicity in bean. *J. Envir. Qual.*, **19**, 359–365.

Kloke, A. (1980). *Richtwerte '80: Orientierungswerte für tolerierbare Gesamtgehalte einiger Elemente in Kulturböden.* VDLUFA Sonderdruck (special print).

Kryger, L. (1980). Differential potentiometric stripping analysis. *Anal. Chim. Acta*, **120**, 19–30.

Kwan, K. H. M., and Smith, S. (1990). Accumulated forms of thallium and cadmium in Lemna minor. II. Relationship between duration of exposure and metal-protein binding. *Chem. Spec. Bioav.*, **3**, 97–103.

Laar, C. V., Reinke, R., and Simon, J. (1994). Determination of thallium in soils by differential pulse anodic stripping voltammetry by means of a mercury film electrode. *Fres. Z. Anal. Chem.*, **349**, 692–693.

Le Blanc, G. A., and Dean, J. W. (1984). Antimony and thallium toxicity to embryos and larvae of fathead minnows (Pimephales promelas). *Bull. Environm. Contam. Toxicol.*, **32**, 565–569.

Lehn, H., and Bopp, M. (1987). Prediction of heavy metal concentrations in mature plants by chemical analysis of seedlings. *Plant Soil*, **101**, 9–14.

Lehn, H., and Schoer, J. (1987). Thallium-Transfer von Böden in Pflanzen. Korrelation zwischen chemischer Form und Pflanzenaufnahme. *Plant Soil*, **97**, 253–265.

Liang Yan-zhong, and Ni Zhe-ming (1994). Atom release of Mn, Co, Ag and Tl in a graphite furnace atomizer with and without palladium modifier. *Spectrochim. Acta*, **49B**, 229–241.

Liem, I., Kaiser, G., Sager, M., and Tölg, G. (1984). The determination of thallium in rocks and biological materials at ng/g levels by differential-pulse anodic stripping voltammetry and electrochemical atomic absorption spectrometry. *Anal. Chim. Acta*, **158**, 179–197.

Lottermoser, B. G., and Schomberg, S. (1993). Thallium content of fertilizers: Environmental implications. *Fres. Environ. Bull.* **2**, 53–57.

Lukaszewski, Z., Ciszewski, A., and Szymanski, A. (1987). *Chem. Anal.*, **32**, 903–912.

Lukaszewski, Z., Pawlak, M. K., and Ciszewski, A. (1980). Determination of thallium and lead in cadmium salts by anodic stripping voltammetry with addition of surfactants to suppress the cadmium peaks. *Talanta*, **27**, 181–185.

Lukaszewski, Z., Zembrzuski, W., and Piela, A. (1996). Direct determination of ultratraces of thallium in water by flow-injection differential-pulse anodic stripping voltammetry. *Anal. Chim. Acta*, **318**, 159–165.

Lukaszewski, Z., and Zembrzuski, W. (1992). Determination of thallium in soils by flow-injection-differential pulse anodic stripping voltammetry, *Talanta*, **39**, 221–227.

Magorian, T. R., Wood, K. G., Michalovic, J. G., Pek, S. L., and Van Lier, M. W. (1974). Water pollution by thallium and related metals, USEPA, Report PB-253 333, Jan.

Makridis, H., and Amberger, A., (1989a). Thallium-Aufnahme aus Zementofenstäuben in Gefäßversuchen mit Grünraps, Buschbohnen und Weidelgras. *Landw. Forsch.*, **42**, 324–332.

Makridis, H., and Amberger, A., (1989b). Wirkung stelgender Thallium-Konzentrationen auf Stoffproduktion und Aufnahme von Thallium und anderen Mineralstoffen durch Buschbohnen und Grünraps in Nährlösungsversuchen. *Landw. Forsch*, **42**, 333–343.

Matthews, A. D., and Riley, J. P. (1970). The occurrence of thallium in sea water and marine sediments. *Chem. Geol.* **6**, 149–152.

Mauras, Y., Premel Cabic, A., Berre, S., and Allain, P. (1993). Simultaneous determination of lead, bismuth and thallium in plasma and urine by inductively coupled plasma mass spectrometry. *Clin. Chim. Acta*, **218**, 201–205.

Minoia, C., Sabbioni, E., Apostoli, P., Pietra, R., Pozzoli, L., Gallorini, G, Nicolaou, G., Alessio, L., and Capodaglio, E. (1990). Trace element reference values in tissues from inhabitants of the European Community. I. A study of 46 elements in urine, blood and serum of Italian subjects. *Sci. Tot. Environ.*, **95**, 89–105.

Nehring, D. (1962). Untersuchungen über die toxikologische Wirkung von Thallium- Ionen auf Fische und Fischnährtiere. *Zeitschrift für Fischerei*, **11**, 557–561.

O'Shea, T. A., and Mancy, K. H. (1976). Characterization of trace metal organic interactions by anodic stripping voltametry. *Anal. Chem.*, **48**, 1603–1607.

Paudyn, A., Templeton, D. M., and Baines, A. D. (1989). Use of inductively-coupled mass spectrometry (ICP-MS) for assessing trace element contamination in blood sampling devices. *Sci. Tot. Environ.*, **89**, 353–352.

Puchelt, H., and Walk, H. (1980). Umweltrelevante Spurenelemente in Böden eines alten Bergbaugebietes. *Naturwissenschaften*, **67**, 190–191.

Qi, W., Chen, Y., and Cao, J. (1992). Indium and thallium background contents in soils in China. *Int. J. Env. Stud.* **40**, 311–351.

Qiao, H., Mahmood, T. M., and Jackson, K. W. (1993). Mechanism of the action of palladium in reducing chloride interference in electrothermal atomic absorption spectrometry. *Spectrochim. Acta*, **48B**, 1495–1503.

Sabbioni, E., Goetz, L., Marafante, E., Gregotti, C., and Manzo, L. (1980). Metabolic fate of different inorganic and organic species of thallium in the rat. *Sci. Tot. Environ.*, **15**, 123–135.

Sager, M. (1993). Determination of arsenic, cadmium, mercury, stibium, thallium and zinc in coal and coal fly ash. *Fuel*, **72**, 1327–1330.

Sager, M. (1992a). Thallium. In *Hazardous Metals in the Environment*, Stoeppler, M. (Ed.). Elsevier, Amsterdam.

Sager, M. (1992b). Speciation of thallium in river sediments by consecutive leaching techniques. *Mikrochim. Acta*, **106**, 241–251.

Sager, M. (1986). *Spurenanalytik des Thalliums*, G. Thieme, Stuttgart.

Sager, M. (1994). Thallium. *Tox. Environm. Chem.*, **45**, 11–32.

Sager, M. (1997). Unpublished results.

Schlee, D. (1986). *Ökologische Biochemie.* Springer, Berlin.

Scholl, W. (1980). Bestimmung von Thallium in verschiedenen anorganischen und organischen Matrices. ein einfaches photometrisches Routineverfahren mit Brillantgrün. *Landw. Forsch.*, **37**, 275–286.

Scholl, W. (1984). Thallium in Tierhaaren. Colloquium Feststoffanalyse Wetzlar 1984, private communication.

Schulze, G., Bonigk, W., and Frenzel, W., (1985). Matrix exchange technique for the simultaneous determination of several elements in flow injection potentiometric stripping analysis. *Fres. Z. Anal. Chem.*, **322**, 255–260.

Shevchenko, L. V., Portretnyi, V. P., and Chuiko, V. T. (1977). Determination of thallium in natural waters by stripping voltammetry after preliminary concentration by coprecipitation (Russ.). *Zh. Anal. Khim.*, **7**, 1448–1450.

Seeger, R., and Gross, M. (1981). Thallium in höheren Pilzen. *Z. Lebensm. Unters. Forsch.*, **173**, 9–15.

Severin, K., Köster, W., and Matter, Y. (1990). *Zufuhr von anorganischen Schadstoffen in Agrarökosysteme mit mineralischen Düngemitteln, Wirtschaftsdüngern, Klärschlamm und Komposten.* VdLUFA Schriftenreihe, Kongreßband, pp. 387–391.

Siedner, G. (1968). Distribution of alkali metals and thallium in some South-West African granites. *Geochim. Cosmochim. Acta*, **32**, 1303–1305.

Siegel, B. Z., and Siegel, S. M. (1976). Effect of potassium on thallium toxicity in cucumber seedlings: Further evidence for potassium-thallium ion antagonism. *Bioinorg. Chem.*, **6**, 341–345.

Solymosi, F., and Bánságy, T., (1971). *Thermal Anal.*, **2**, 289–299. *Proceedings 3rd ICTA Davos 1971*, Birkhäuser Verlag, Basel.

Sprung, S., Rechenberg, W., and Bachmann, G. (1994). Umweltverträglichkeit von Zement. *Zement Kalk Gips*, **47**, 456–461.

Sturgeon, R. E., Willie, S. N., Zheng, J., Kudo, A., and Gregoire, D. C. (1993). Determination of ultratrace levels of heavy metals in arctic snow by electrothermal vaporization inductively coupled plasma mass spectrometry. *J.A.A.S.*, **8**, 1053–1057.

Tessier, A., Campbell, P. G., and Bisson, M. (1979). Sequential extraction procedure for the speciation of particulate trace metals. *Anal. Chem.*, **51**, 844–851.

Tilch, J., Hoffmann, C., and Lüdke, Ch. (1996). Air analysis by ETA-LEAFS and ETV-ICP-MS. Poster at the ISEAC in Vienna 1996 (pers. commun.).

Török, P., and Schmall, W. (1982). Umweltschadstoff Thalliumsulfat. *Dtsch. Ärztebl.*, **79**, 31–32.

Tremel, A., Masson, P., Sterckeman, Th., Baize, D., and Mench, M. (1995). Preliminary investigation on thallium in soils in France. 3rd Int. Conf. Biogeoch. Trace Elem., Paris, May.

Tsakovski, S., Ivanova, E., and Havezov, I. (1994). Flame AAS determination of thallium in soils. *Talanta*, **41**, 721–724.

Valerio, F., Brescianini, C., Mazzucotelli, A., and Frache, R. (1988). Seasonal variation of thallium, lead and chromium concentrations in airborne particulate matter collected in an urban area. *Sci. Total Environ.*, **71**, 501–509.

Völkening, J., and Heumann, K. G. (1990). Heavy metals in the near-surface aerosol over the Atlantic Ocean from 60° South to 54° North. *J. Geophys. Res. [Atmos.]*, **95**, 20623–20632.

Waidmann, E., Emons, H., and Dürbeck, H. W. (1994). Trace determination of Tl, Cu, Pb, Cd, and Zn in specimens of the limnic environment using isotope dilution mass spectrometry with thermal ionization. *Fres. J. Anal. Chem.*, **350**, 293–297.

Waidmann, E., Hilpert, K., and Stoeppler, M. (1990). Thallium determination in reference materials by isotope dilution mass spectrometry (IDMS) using thermal ionization. *Fres. J. Anal. Chem.*, **338**, 572–574.

Waidmann, E., Stoeppler, M., and Heininger, P. (1992). Determination of thallium, in sediments of the River Elbe using isotope dilution mass spectrometry with thermal ionization. *Analyst,* **117,** 295–298.

Wang, J., and Jianmin, Lu, (1993). Adsorptive stripping voltammetry of trace thallium. *Anal. Chim. Acta,* **282,** 329–333.

Wegener, S., Pohl, B., Folz, M., and Jansen, A. (1992). Determination of thallium in environmental materials with graphite furnace AAS. A comparison of five different graphite tube types. *Fres. J. Anal. Chem.,* **343,** 153.

Welz, B., Schlemmer, G., and Mudakavi, J. R. (1988). Investigation and elimination of chloride interference on thallium in graphite furnace atomic absorption spectrometry. *Anal. Chem.,* **60,** 2567–2572.

Yoshida, S., Muramatsu, Y., Tagami, K., and Uchida, S. (1996). Determination of major and trace elements in Japanese rock reference samples by ICP-MS. *Int. J. Environ. Anal. Chem.,* **63,** 195–206.

Yoshida, S., Muramatsu, Y., and Ban-nai, T. (1996). Accumulation of radiocesium and trace elements in mushrooms collected from Japanese forests. *Mitt.d. Österr. Bodenkundl. Ges.,* **53,** 251–258.

You, N., and Neeb R. (1983). Verbesserung der invers-voltammetrischen Bestimmung des Thalliums neben Blei durch elektrochemische Maskierung. *Fres. Z. Anal. Chem.,* **314,** 394–397.

Zheng, Y., and Wang, Y. (1995). Applications of analyses on the basis of characteristic mass: Determinations of indium, silver, and thallium in drainage sediment and geochemical samples. *Talanta,* **42,** 361–364.

Zitko, V. (1975). Toxicity and pollution potential of thallium. *Sci. Tot. Environ.,* **4,** 185–192.

Zitko, V., Carson, W. V., and Carson, W. G. (1975). Thallium: Occurrence in the environment and toxicity to fish. *Bull. Environ. Contam. Toxicol.,* **13,** 23–30.

6

DETERMINATION OF THALLIUM BY ELECTROANALYTICAL TECHNIQUES

Miquel Esteban, Cristina Ariño, and José Manuel Díaz-Cruz

Department of Analytical Chemistry, University of Barcelona, Av. Diagonal 647, 08028 Barcelona, Spain

Thallium in the Environment, Edited by Jerome O. Nriagu.
ISBN 0-471-17755-5. © 1998 John Wiley & Sons, Inc.

6.6. Summary
References

6.1. INTRODUCTION

Thallium is a toxic heavy metal that causes both chronic and acute poisoning. It can be found in minerals such as crookesite [(CuTlAg$_2$)Se] and lorandite (TlAsS$_2$), and at trace levels in pyrites and zinc blends. It is found at higher concentrations (0.02–1.5 mg/kg) in silicate and other rocks. Average concentration in Earth's crust ranges between 0.3 and 0.5 μg/g. Thallium is also introduced into the environment mainly as waste from the production of cement, lead, zinc, and cadmium and by coal combustion (Khopkar and Gandhi, 1995). Flue dust from burners used with thalliferrous pyrites can contain up to 0.25% (w/w) thallium, while particulate matter collected in bags (to prevent pollution due to suspended particles) from factories where thalliferrous lead–copper ores are roasted can contain up to 0.5%. Thallium and its compounds are used in alloys with Pb, Cd, Sn, and Bi (usually <1%) and in optical systems, as well as in chemical synthesis.

Thallium is also encountered in various biological samples such as plant material, bone, and animal organs at concentrations ranging from 0.01 to 5 mg/kg, while traces of thallium are also found in fish.

In aqueous solution thallium can be present in two different redox states: Tl(I) and Tl(III). The octahedrical ion Tl(H$_2$O)$_6^{3+}$ behaves as an acid (pK_a $\approx$ 1.15), and the solutions of Tl(III) salts are hydrolyzed yielding TlOH^{2+} and the colloidal hydrated oxide, even at the pH range 1–2.5. Trivalent thallium can be stabilized in solution by formation of the successive complexes with chloride up to TlCl$_6^{3-}$, the complex TlCl$_2^+$ being especially stable, as stated in Cotton and Wilkinson (1966). In general, Tl(III) exhibits properties analogous to those of compounds of zinc and mercury.

In aqueous solutions Tl(I) is more stable than Tl(III):

$$Tl^{3+} + 2e \leftrightarrows Tl^+ \qquad E^\circ = 1.25 \text{ V}$$
$$E^{\circ\prime} = 0.77 \text{ V (1 } M \text{ HCl) and 1.26 V (1 } M \text{ HClO}_4)$$

Monovalent thallium exhibits properties similar to those of silver and lead and can also be reduced to the metal:

$$Tl^+ + 1e^- \leftrightarrows Tl(s) \qquad E^\circ = -0.336 \text{ V}$$
$$E^{\circ\prime} = -0.55 \text{ V (1 } M \text{ HCl) and } -0.33 \text{ V (1 } M \text{ HClO}_4)$$

Under redox potential, E_H, and pH conditions usually found in natural waters, Tl(I) is considered to be the predominant thallium species. Thus, speciation models for seawaters indicates that, for Tl(I), 53% is present as aqueous complex, 44% as chlorocomplexes, 2% as sulfatocomplexes, and less than 1% as hydroxy- and fluorocomplexes. According to Whitfield and Turner, (1981), Tl(III) is only present as hydroxocomplexes (precipitated or/and colloidal hydrated oxide).

In general, the total thallium content in unpolluted environmental samples is at the nanogram per gram level. In sea water, free Tl(I) has been reported to be at the level of ca. 10^{-9} M (Zirino, 1981), but concentrations in sea water and in rain have been reported to be less than 5×10^{-10} M, as discussed later.

Since the concentration of thallium present in environmental compartments is low, electroanalytical techniques may be suitable for its determination. The purpose of this chapter is to summarize recent work in this regard. The more general aspects of the various techniques are briefly reviewed for the nonspecialists.

6.2. POTENTIOMETRY

Potentiometry is an electroanalytical technique that allows the determination of an ion in solution by measuring the potential of the cell formed by an indicator electrode, which responds to that ion, and a reference electrode. Different kinds of electrodes that respond as selectively as possible to certain ions have been developed and are known as ion-selective electrodes (ISE). The glass electrode, widely used for the determination of pH, remains the most familiar example of an ISE, but nowadays a great number of different ISEs are available, as shown in several reviews by Koryta (1972, 1977, 1979, 1982, 1984, 1986, 1988, 1990) and by Freiser (1978). In general, despite their low sensitivity, the use of ion-selective electrodes provides good methods to determine ions using very simple instrumentation.

Ion-selective electrodes can be regarded as electrochemical half-cells consisting of a suitable reference electrode/reference solution system, separated from the sample by a membrane (Fig. 1).

The nature of the membrane determines the ions to which an electrode responds. Two types of membrane are usually employed: (i) liquid membranes, which are porous plastic membranes holding a liquid ionic exchanger whose nature depends on the analyte to determine, and (ii) solid membranes prepared from cast pellets of salts or from a slice of a crystal of a certain salt, usually doped with another salt to improve its conductivity.

When an ISE is placed in a solution, an electric potential is developed across the membrane, the magnitude of which is dependent on the activity of the ion to which the electrode responds. This potential is measured by completing the electrochemical cell by means of a suitable reference electrode [usually a saturated calomel electrode (SCE) or a Ag/AgCl electrode], and connecting the electrodes to a high-impedance millivoltmeter. Under normal

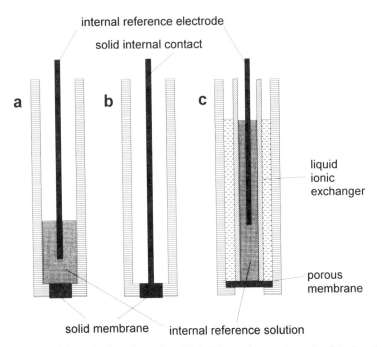

Figure 1. Types of ion-selective electrodes: (a) Solid–membrane electrode, (b) all solid-state solid–membrane electrode, (c) liquid–membrane electrode.

conditions of use, the potential E of the cell is ideally given by the Nernst equation.

Some years ago Tl(I) was determined by potentiometric titration, instead of direct potentiometry, using inert redox indicator electrodes (e.g., Pt). In these titrations, redox reagents such as potassium permanganate, potassium chromate, or potassium dichromate (in the presence of iodine monochloride and carbon tetrachloride) were used. Further, the development of different kinds of selective membranes allowed the direct determination of Tl(I) in the samples. Both types of membrane (liquid and solid) have been used in the determination of Tl(I).

Coetzee and Basson (1977) described a Tl(I)-sensitive liquid membrane based on a hydrophobic anionic substance, such as thallium tetra(*m*-trifluoromethylphenyl)borate, that allowed determination of Tl(I) concentrations as low as 10^{-5} M in the pH range 3–9. Further, Coetzee (1985) described a similar membrane in which the organic phase is a saturated solution of Tl(I) cyanotriphenylborate in 4-ethylnitrobenzene. The advantage of this electrode is the simplicity of the preparation of the active material and hence the electrode. Szczepaniak and Ren (1976, 1980) proposed as active reagent in the liquid membrane the covalent salt of Tl(I) with *o,o'*-

didecyldithiophosphoric acid. The electrode has nernstian behavior in the pTl range 1–5.5. Alkali and alkaline earth metal cations do not interfere (selectivity coefficients, $K_{Tl/M} < 10^{-5}$) and the potentials are unaffected by pH in the range 5–12. Ethylene-diaminetetraacetic acid (EDTA) and NaCN masked almost all the mono- and bivalent cations that interfere in the measurement. The use of crown ethers in Tl(I) liquid–membrane electrodes has been proposed by various authors. Tamura et al. (1980) studied three derivatives of the bis(benzo-15-crown-5) (molecule I in Fig. 2, where $n = 3, 5, 7$) and a monocyclic crown ether, the hexanoyloximethyl-benzo-15-crown-5 (molecule II in Fig. 2). The crown ethers were immobilized in a polyvinyl chloride (PVC) membrane with o-nitrophenyloctylether or dipentylphthalate as plasticizer. The nernstian responses of the different electrodes are in the pTl range 2–5. Huang et al. (1984) prepared a PVC membrane containing another derivative of the bis(benzo-15-crown-5) (molecule III in Fig. 2) to determine Tl(I) in the range 2×10^{-6}–10^{-2} M. The selectivity coefficients for the interferences are in the order $K^+ > Ag^+ > Na^+ > NH_4^+$.

Thiacrown ethers (macrocyclic polythiaethers) have also been the focus of research because of their properties as selective complexing agents. They are typically uncharged compounds with an internal sulfur-donor-rich cavity capable of soft cation encapsulation with selectivity. For determination of Tl(I), Masuda et al. (1994) studied the performance of ISEs with a plasticized PVC membrane containing one of the five macrocyclic thiaethers shown in Figure 3. Among them only 1, 4, 8, 11-tetrathiacyclotetradecane plasticized by dioctylphthalate showed good performance for Tl(I). Electrodes containing the other macrocyclic thiaethers did not show nernstian response to Tl(I) activity.

Electrodes containing crown ethers and thallium(I) tetraphenylborate have also been prepared and studied. Yang et al. (1987) proposed the use of 4-methyldibenzo-30-crown-10 and thallium(I) tetraphenylborate. The resulting electrode shows nernstian response in the 5×10^{-6} – 10^{-2} M Tl(I) range, showing high selectivity, stability and reproducibility.

The use of various dyes has been proposed by Fogg et al. (1975) and Dan and Dong (1988). $TlCl_4^-$ salts of Sevron Red L (Color Index Basic Red 17), Sevron Red GL (C.I. 11,085), flavinduline O (C.I. 50,000), phenazinduline O and butylrhodamine B have been proposed as electroactive agents.

A variety of solid-state ion-selective electrodes for thallium have been investigated, which are based on miscellaneous precipitates embedded in plastic matrices. This group includes electrodes with Tl(I) molybdo- or tungstophosphate (Coetzee and Basson, 1973), both in epoxy matrix, electrodes based on tungstoarsenates in Araldite (Malik et al., 1976) or membranes of α-picolinium molybdoarsenate in Araldite (Jain et al., 1980).

Fombon et al. (1979) and Vlasov et al. (1981) described the use of polycrystalline membranes constructed with Tl_2S/conductive matrix or Tl_4HgI_6/conductive matrix mixtures and HgI_2/TlI polycrystals, respectively. The use of TlI/AgI has also been described in the book of Vesely et al. (1978).

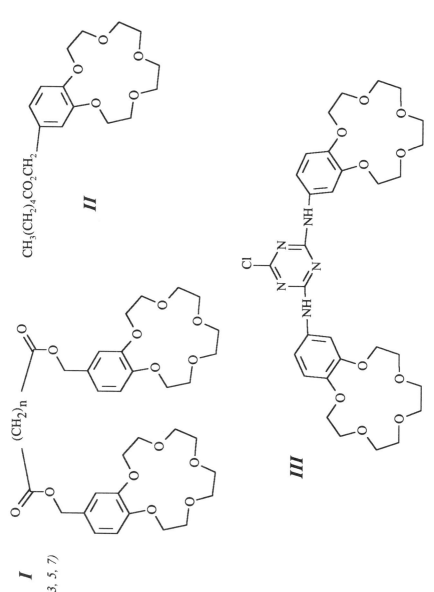

Figure 2. Structure of some crown ethers used in the development of ISEs for Tl(I).

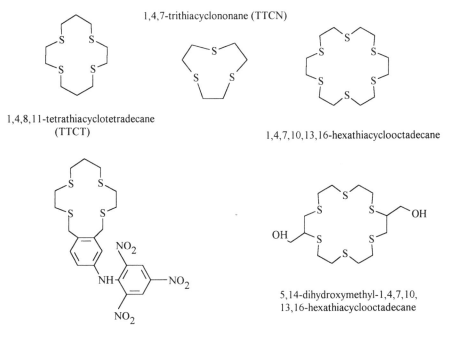

1,4,7-trithiacyclononane (TTCN)

1,4,8,11-tetrathiacyclotetradecane
(TTCT)

1,4,7,10,13,16-hexathiacyclooctadecane

NO_2

NH—NO_2

NO_2

4'-picrylaminobenzo-1,4,8,11-tetrathiacyclopentadec-13-ene

OH

OH

5,14-dihydroxymethyl-1,4,7,10,
13,16-hexathiacyclooctadecane

Figure 3. Structure of some thiacrown ethers used in the development of ISEs for Tl(I).

The glass electrode can also be used for Tl(I) determination. Depending on the composition of the glass membrane, the response may be chiefly to H^+ ion, or it may be to other monovalent cations, including Tl(I), some divalent cations or even to organic cations. Eisenman (1965) pointed out that most glass electrodes that are responsive to alkali cations respond rapidly and reproducibly to Tl(I), with a nernstian slope for standard Tl(I) solutions. The Tl(I) selectivity of glass electrodes is, moreover, a systematic function of glass composition, which can be expressed in terms of Na^+–K^+ selectivity, and knowledge of the Na^+–K^+ selectivity of a given glass suffices to define the approximate selectivity for Tl(I). The recommended composition of glass electrodes for Tl(I) determination are KAS 20-5B or KAS 20-4B.

6.3. POLAROGRAPHY

6.3.1. Introduction

IUPAC (Freiser and Nancollas, 1987) has recommended that the term polarography be reserved for the measurement of current–voltage curves at a liquid electrode whose surface is continuously renewed, while the term voltammetry should be used only for the measurement of current–voltage at solid or stationary electrodes.

Basically, polarographic techniques consist of measuring the current that flows in a solution between a reference electrode and a dropping mercury electrode (DME) as a function of the potential difference applied externally. In most modern systems, however, more accurate measurements are made with a three-electrode system: The current is measured between the DME and an auxiliary electrode, and the potential is monitored between the DME and the reference electrode.

Figure 4 shows a typical experimental setup for polarography. The DME is a glass bar with a capillary channel through which mercury flows under gravity or propelled by other mechanical means (gas pressure, knocking, etc.). This produces small mercury drops that grow and fall down, to be replaced by a new drop. In fact, recent advances in instrumentation have led to the static mercury drop electrode (SMDE), an improved version of DME, in which the drops grow rapidly and maintain the same size. The low surface area of the mercury drops in both DME and SMDE guarantees that all the electrochemical processes taking place on them affect only a very thin layer

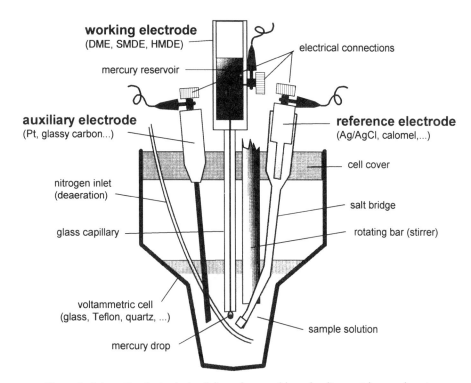

Figure 4. Schematic of a typical cell for polarographic and voltammetric experiments.

(the diffusion layer). This means that the composition of the bulk solution remains unchanged.

Polarographic techniques are carried out without solution stirring and in the presence of an inert electrolyte (KNO_3, $NaClO_4$, diluted mineral acids, etc.) that increases the conductivity of the solution without suffering any electrochemical reaction. In this way, among the three possible ways of mass transport toward the electrode (convection, ionic migration, and diffusion) only the diffusion has to be considered, which simplifies considerably the interpretation of the signals obtained.

Figure 5 shows the principles of several polarographic techniques, applied to a substance that can be reduced on the DME. In direct current polarography (DCP) a different potential is applied to each mercury drop and the current is measured during the full drop life (classical DCP) or, better, at a fixed time at the end of the drop (Tast DCP). The plot of the currents (I) as a function of the potentials (E) constitutes the direct current (dc) polarogram, in which the number of points corresponds to the number of drops used. The sigmoidal shape can be explained by assuming that too positive potentials do not produce reduction of the species (zero current). More negative potentials (i.e., more reducing) produce an increasing negative current due to the more favorable reduction on the electrode (the process is controlled by the electrochemical reaction). Finally, still more negative potentials produce a constant current (limiting current) because the process is controlled by the diffusion of the species toward the electrode, which cannot increase any further. The value of the limiting current (I_{lim}) is directly proportional to the concentration of the species, which can be used for its determination. Another important parameter is the half-wave potential $E_{1/2}$ (potential for which $I = I_{lim}/2$). This parameter is characteristic of the different species and allows their identification. This also allows the simultaneous determination of several species in a single potential sweep if their electrode processes take place at sufficiently different potentials to yield separate signals.

In techniques like alternating current polarography (ACP) or differential pulse polarography (DPP), the potential excitation pattern is more sophisticated and the signal obtained for each drop involves different current measurements. For instance, in DPP the signal is the difference between the currents measured after and before the application of a potential pulse. Depending on the potential excitation pattern and on the way the current is measured, the polarogram can have different shapes. In the case of ACP and DPP, the signals are peak shaped, with a height (I_{peak}) proportional to the species concentration and a peak potential (E_{peak}) characteristic of the substance. Other polarographic techniques (not shown in Fig. 5) are normal pulse polarography and reverse pulse polarography, both producing sigmoidal signals. For more details we refer to Bond (1980).

A more recent technique is square-wave voltammetry (SWV), which is a type of pulse polarography that offers the advantage of great speed and high sensitivity. An entire voltammogram is obtained in a few seconds or less.

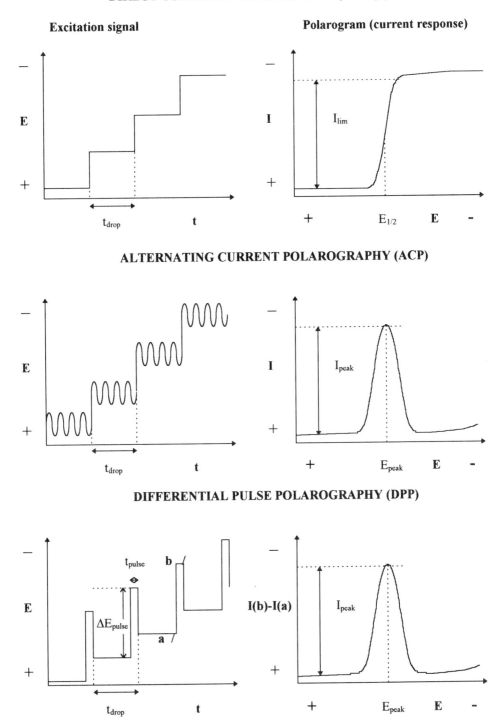

Figure 5. Potential excitation signals and current responses for some polarographic techniques.

A complete description of the technique can be found in Osteryoung and Osteryoung (1985). It can be performed in a DME, an SMDE, or a hanging mercury drop electrode (HMDE), which will be further described in more detail. This is why this voltammetric technique is usually included among the polarographic techniques.

From an analytical point of view, DPP and SWV are surely the most suitable techniques, because of the very high signal-to-noise ratio obtained. However, other techniques can also be useful depending on the experimental conditions (concentration level, adsorption of the species on the mercury drop) and the kind of information required about the system (concentration, electrochemical reversibility, kinetic parameters).

6.3.2. Applications

Studies carried out by classical polarography have shown that, in aqueous solutions, Tl(I) produces a well-defined cathodic wave at the DME, corresponding to the reversible reaction:

$$Tl^+ + 1e^- \rightleftharpoons Tl(Hg)$$

In this equation, the term Tl(Hg) means that the reduced Tl^0 forms an amalgam with the mercury on the drop. This process is favored by the high solubility of Tl^0 in mercury [42% w/w as shown in Vydra et al. (1976)]. The limiting current of such a wave was found to be linearly related to the thallium concentration. The half-wave potential ($E_{1/2}$) appeared to be close to -0.5 V (vs. SCE) and quite insensitive to the presence of strong complexing agents as a consequence of the poor ability of Tl(I) to form complexes.

These properties were applied to the polarographic determination of thallium. However, this determination is subject to interference from certain metal ions that are reduced in the same potential region. These metal ions include lead, cadmium, copper, bismuth, indium, titanium, and iron. In most samples Tl is found simultaneously to other more frequent trace metals such as Pb and Cd. In the case of Pb, for instance, the determination of Tl(I) is hindered by the small peak (or half-wave) potential separation between Tl(I) and Pb(II), which depends on the medium used but is ca. 70 mV in the absence of strong complexing agents.

A strategy for resolving overlapping voltammetric waves or peaks was to use chemical reagents for complexing one component in the mixture, in order to separate the half-wave or peak potential values or to mask one component completely. One of the first ligands used for this purpose was EDTA (Pribil et al., 1951). However, the ligands often shift waves that are initially at more positive potentials toward the region of the thallium wave.

Additional selectivity was obtained with surface-active agents that proved to be able to inhibit some electrochemical processes. This phenomenon was termed "electrochemical masking" by Reilley et al. (1956), who observed that the addition of gelatin decreased or even suppressed the reduction waves of several metal ion–EDTA complexes without causing appreciable distortion of the reduction waves of the corresponding free metal ions. Further studies discovered similar effects for other ligands and surface-active agents. As Tl(I) is hardly complexed, its signal is not markedly affected by the addition of these substances.

Later, methods were proposed that combined the effects of a strong complexing agent and a surface-active substance in order to eliminate most of the interfering signals. For instance, Shetty et al. (1962) used sodium triphosphate 0.1 M and 0.1% camphor to suppress the interference of Cu, Bi, Pb, and Fe. The method, however, was reported to be quite pH sensitive. Jacobsen and Kafland (1963) proposed a medium containing diethylenetriaminepentaacetic acid (DTPA) 10^{-2} M, 0.01% of Triton X-100 and acetate buffer 0.2 M at pH 4–5. Under these conditions, interferences of Co, Ni, Zn, Cu, Pb, Bi, Cd, and Cr were absent and only high concentrations of Ag and Fe could cause problems.

Due to the low sensitivity of classical DCP, these methods were applicable only to solutions containing 10^{-4}–10^{-3} M thallium. With the introduction of pulse polarography, this method (complexation + electrochemical masking) was adapted to the new techniques (especially DPP and SWV), resulting in an improvement in the detection limits for thallium.

Hoeflich et al. (1983) used DPP to determine Tl^+ and $(CH_3)_2Tl^+$ simultaneously at pH 4, 7, and 10 (acetate and phosphate buffers) with average detection limits of 130 ppb for Tl^+ and 250 ppb for $(CH_3)_2Tl^+$, respectively. For these species, the electrochemical reactions assumed were:

$$Tl^+ + 1e^- \rightleftharpoons Tl(Hg) \quad \text{(reversible)}$$

$$(CH_3)_2Tl^+ + 3e^- + 2H^+ \rightarrow 2\ CH_4 + Tl(Hg) \quad \text{(irreversible)}$$

These reactions produce two well-defined peaks at -0.5 and -1.1 V (vs. SCE), respectively, although the last peak is strongly pH dependent. The interferences of Pb(II), Zn(II), and Cd(II) were eliminated with EDTA, as shown in Figure 6.

Wilgocki and Cyfert (1989) used a medium containing ethanediamine 2 M, NaOH 0.25 M, $NaClO_4$ 0.6 M, and 0.01% of gelatine to achieve complete separation of the DPP peaks of Tl(I), Pb(II), and Cd(II). However, the method was conceived not for the determination of thallium but for the analysis of lead.

Nevertheless, other approaches for signal separation have been proposed. Among them, the methods based on purely numerical curve resolution techniques were better accepted. The test solution most widely used to assess the resolution of the various methods is a mixture of Pb and Tl at varying proportions, because of the signal overlap. To the best of our knowledge, no application of these methods to nonsynthetic samples is available. A few recent

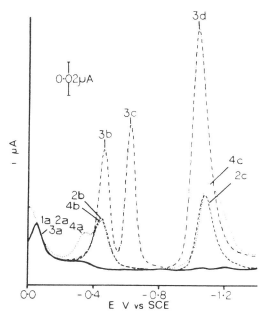

Figure 6. DPP of mixture TlNO$_3$ and (CH$_3$)$_2$ TlNO$_3$ in the presence of interfering metals and EDTA, acetate buffer pH 7: (–) blank, (———) 4 ppm TlNO$_3$ and 4 ppm (CH$_3$)$_2$ TlNO$_3$, (·—·—·—) 4 ppm Pb^{2+}, Cd^{2+}, Zn^{2+} added, (···) solution made 0.01 M EDTA; (A peak) Cu^{2+}, (B peak) Tl$^+$, (C peak) Cd^{2+}, (D peak) (CH$_3$)$_2$ Tl$^+$ and Zn^{2+}, (E peak)(CH$_3$)$_2$ Tl$^+$, (F peak) (CH$_3$)$_2$ Tl$^+$, Pb^{2+}, Cd^{2+}, and Zn^{2+}. (Reproduced from *Anal. Chem.* **55,** 1591–1595 (1983), with the permission of American Chemical Society.)

examples are summarized in what follows. For previous investigations, see the literature cited.

Pižeta (1994) used deconvolution involving fast Fourier transformation (FFT) to solve nonresolved DPP and SWV peaks of Tl and Cd at different ratios at the level of 10^{-4}–10^{-5} M. Later, Pižeta et al. (1994) studied three ways of data pretreatment that can be used as part of a deconvolution procedure of DPP signals from synthetic mixtures of Pb and Tl or of simulated DPP polarograms. From this investigation, some recommendations are given according to the shape of the signal treated. More recently, Grabarić et al. (1995) proposed a signal ratio method for resolution enhancement in DPP and related techniques. The method proposed was able to resolve Pb from at least 40-fold excess of Tl, and Tl from at least 20-fold excess of Pb at micromolar concentration range with reproducibility better than 5%.

Besides the determination of thallium, another application of polarography is the study of the macrocyclic complexes of this metal. The ligands able to form such complexes (macrocyclic ligands) have a cyclic structure around the complexation site, so the site acts as a cavity in which only the ions having the proper size can be introduced. That makes these ligands very selective with respect to the size of the metal to be complexed, which can be used, for instance, to remove harmful cations, without affecting the concentration of the biologically important ones (like Na^+, K^+, Mg^{2+}, Ca^{2+}, Zn^{2+}).

Polarographic methods have been used to determine the stability constants of a large number of macrocyclic metal complexes in aqueous and nonaqueous media. In the case of thallium, Koryta and Mittal (1972) studied by DCP its complexes with dibenzo-18-crown-6 and dicyclohexyl-18-crown-6, in both water and methanol, Khalil et al. (1985) determined by ACP and DCP the stability constants in water of the complexes of Tl(I) with 15-crown-5, benzo-15-crown-5, 18-crown-6, and dicyclohexyl-18-crown-6, and Parham and Shamsipur (1991) studied by DPP the complexation reactions between Tl^+, Cd^{2+}, Hg^{2+}, Pb^{2+}, and 18-crown-6, 1,10-diaza-18-crown-6 (C22), and cryptand (C222) in acetonitrile–water mixtures. Figure 7 summarizes the structure of all these macrocyclic ligands.

6.4. STRIPPING VOLTAMMETRY

6.4.1. Introduction

Due to the low concentration levels of thallium in the environment, polarographic techniques do not allow its determination in unpolluted natural samples. Then, more sensitive techniques are necessary.

Stripping voltammetry (SV) comprises a variety of electroanalytical approaches having a preconcentration step, usually onto a mercury electrode, prior to the voltammetric measurement. This preconcentration step can be viewed as an effective electrochemical extraction in which the analyte is preconcentrated in the mercury phase to a much higher level than that existing in solution. The combination of the preconcentration step with the electrochemical measurement of the concentrated analyte generates an extremely favorable signal-to-background ratio that characterizes stripping voltammetric techniques. For more information about these techniques we refer to Bond (1980), Vydra et al. (1976), Wang (1985), and Barendrecht (1967).

Among the different possibilities of SV, anodic stripping voltammetry (ASV) is the most popular approach. Anodic stripping voltammetry consists of three steps. First, a constant deposition potential is applied for a fixed deposition time, while the solution is stirred. In this deposition step, a small portion of the analyte is reduced onto the electrode surface undergoing the corresponding amalgamation [formation of M(Hg)]. In the second step, the stirring is stopped and the deposition potential is maintained for a fixed time. Finally, the third step is a potential sweep in the anodic direction that produces an electrochemical reoxidation of the amalgam to the metal ion (stripping step). This potential sweep can be carried out according to different patterns.

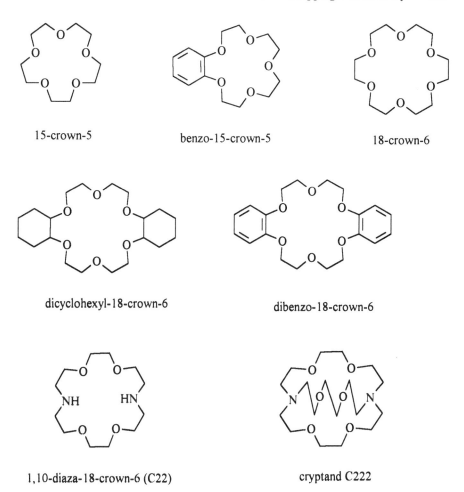

15-crown-5 benzo-15-crown-5 18-crown-6

dicyclohexyl-18-crown-6 dibenzo-18-crown-6

1,10-diaza-18-crown-6 (C22) cryptand C222

Figure 7. Crown ethers whose stability constants with Tl(I) have been determined polarographically.

In the simplest case, the potential changes linearly with time. Furthermore, potential pulses similar to those used in DPP, ACP, or SWV can be applied to the electrode where M(Hg) was formed, in order to improve the signal-to-noise ratio. Then, differential pulse anodic stripping voltammetry (DPASV), alternating current anodic stripping voltammetry (ACASV), and square wave anodic stripping voltammetry (SWASV), respectively, are performed. Figure 8 shows a simplified scheme of the differents steps of ASV, while Table 1 summarizes the electrochemical processes taking place.

The preconcentration step makes stripping analysis much more sensitive than direct polarographic techniques, because the concentration of the

Applied potential vs. time   **Voltammogram (current vs. potential)**

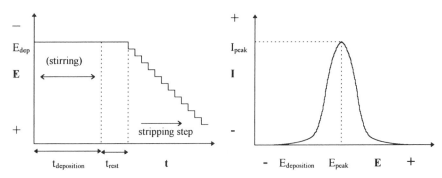

Figure 8. Sequence for stripping voltammetry measurements.

metal in the mercury can be 10^2–10^3-fold greater than that of the metal ion in the sample solution, and it can be improved by increasing the deposition time.

The global electrochemical process produces a peak-shaped voltammogram that provides analytical information since: (i) the peak area and the peak current are proportional to the metal concentration in the sample solution and can be used for quantitative determinations, if reproducible deposition conditions are attained; and (ii) the peak potential serves to identify the metals.

In ASV mercury electrodes are mainly used: HMDE and mercury film electrode (MFE). Hanging mercury drop electrode is able to produce a mercury drop that hangs at the tip of the capillary during the whole ASV experiment. In contrast, MFE consists of a glassy carbon surface where a thin film of Hg is electrolytically deposited prior to the ASV measurement. In general, HMDE provides more reproducible results than MFE, whereas MFE is more sensitive and presents better resolution.

Some other variants of SV have been developed. We summarize only those that have been used for Tl determination (Table 1).

Adsorptive stripping voltammetry (AdSV) is a modality of SV also consisting of three steps (Paneli and Voulgaropoulos, 1993). In the first step, a constant deposition potential is applied for a fixed deposition time, while the solution is stirred. In this step, a small portion of the analyte is adsorbed onto the electrode surface, but it usually does not undergo electrochemical transformation. In the second step, the stirring is stopped and the deposition potential is maintained for a fixed time. Finally, the third step is a potential sweep that produces an electrochemical oxidation or reduction of the species adsorbed on the electrode (Table 1).

In many cases the analyte does not adsorb by itself onto the electrode, and so it must be transformed into a derivative with such an ability. This is the

Table 1   Different Techniques of Stripping Analysis

Preconcentration Step	Measuring Step
Schemes of Electrolytic Preconcentration	
Anodic stripping voltammetry (ASV)	
$M^{n+} + ne \rightarrow M(Hg)$	$M(Hg) \rightarrow M^{n+} + ne$
Potentiometric stripping analysis (PSA)	
$M^{n+} + ne \rightarrow M(Hg)$	$M(Hg) + Oxid. \rightarrow M^{n+} + ne$
$M^{n+} + X^{m-} \rightarrow M_mX_n + ne$	$M_mX_n + Red. \rightarrow M + X^{m-}$
Schemes of Nonelectrolytic Preconcentration	
Adsorptive stripping voltammetry (AdSV) onto conventional electrodes	
$M^{n+} + L^{m-} \rightarrow (M_mL_n)_{ads}$	$(M_mL_n)_{ads} \rightarrow M^{n+} + L^{m-}$
	$(M_mL_n)_{ads} \rightarrow M^{n+} + L^{p-} + (m-p)e$
	$(M_mL_n)_{ads} + ne \rightarrow M + L^{m-}$
Stripping voltammetry with preconcentration on chemically modified electrodes	
$M^{n+} + R(LH)_n \rightarrow R(LM)_n + nH^+$	$R(LM)_n + me \rightarrow M^{(n-m)+}$
$Org^{n+} + R(LH)_n \rightarrow$ $R(LOrg)_n + nH^+$	$R(LOrg)_n + me \rightarrow Org^{(n-p)+}$
Abrasive stripping voltammetry (AbrSV)	
Mechanical transfer of A	$A + e \rightarrow A^-$

case for most metal ions, which are usually analyzed in the form of organic complexes [e.g., Ni(II)-dimethylglyoxyme]. Moreover, the deposition potential must be chosen carefully in order to prevent electrochemical processes and allow effective adsorption of the analyte (e.g., a positive potential enhances the adsorption of negatively charged species due to electrostatic attraction). It is also important to notice that the amount of substance adsorbed (and thus, the stripping signal) is proportional to both the deposition time and the bulk concentration of analyte as long as the surface is not saturated. As a consequence, linearity must be tested in the calibration plots.

Although most SV applications are related to the use of mercury electrodes, the development and use of chemically modified electrodes (CME) is an active subject in electroanalytical chemistry. Chemically modified electrodes have been deliberately treated with some reagent, having desirable properties, so as to take on the properties of the reagent (Arrigan, 1994). Chemically modified electrodes provide an approach to the development of analytical procedures employing immobilized reagents, and many methods have been described for their preparation (Murray, 1984).

It is therefore possible to design electrodes for various purposes and applications. One is to preconcentrate an analyte so that very low concentrations may be measured in another medium. In this case, SV methods can be applied to CMEs using similar schemes to those mentioned above. The first step is the preparation of the CME. Further, a deposition step is performed to accumulate the analyte on the electrode surface, by complexation or adsorption. Later, the voltammetric scan provides the current data needed for the analyte quantitation. For more details see the references by Arrigan (1994) and Murray (1984).

Abrasive stripping voltammetry (AbrSV) is a recent technique of potential usefulness in solid-state analysis (Scholz and Lange, 1992). It is based on the mechanical transfer of trace amounts of a solid sample onto the surface of an appropriate solid electrode (deposition step) and a further oxidation or reduction of the analyte by integration of the electrode in an electrochemical cell (stripping step).

The next sections illustrate the applications of SV to the thallium determination.

6.4.2.   Anodic Stripping Voltammetry

Because thallium can easily form an amalgam with mercury, ASV seems to be a good technique for its determination. The application of ASV shows that Tl(I) gives a well-defined peak in both HCl (Neeb and Kiehnast, 1968) and pH 4.6 acetate buffer (Batley and Florence, 1974) electrolytes. At an MFE the peak occurs at -0.67 V (vs. SCE), while the peak is at HMDE at -0.47 V. The heights of these peaks are linearly related to the thallium concentration in a wide concentration range.

Klahre et al. (1978) determined simultaneously Cd(II), Cu(II), Pb(II), and Tl(I) in drinking water by DPASV in a 0.3–0.5 M NaOH pH $>$ 12 medium. They applied a deposition potential of -1 V (vs. SCE) to an HMDE for 5 min. Under these conditions it is easy to measure 1 μg/kg. The application of the method to some drinking waters gave levels above 1 μg Tl/kg. Dieker and van der Linden (1975) determined Tl(I) in the presence of 10^3-fold excess of lead and copper by means of DPASV in a thin mercury film electrode in a medium of KNO_3 at pH below 3.5–4.0. However, in general, as in polarographic analysis, stripping voltammetric determination of thallium is subject to interferences from different metal ions, including lead, cadmium, copper, bismuth, indium, titanium, and iron, which are reduced in the same potential region as Tl(I).

Methods based on purely numerical curve resolution techniques have been proposed for solving those interferences. As mentioned in the polarographic section, the mixtures of Pb and Tl at different concentration ratios are probably the most widely used. As an example, Henrion et al. (1990) applied partial least-square (PLS) regression for signal resolution of DPASV peaks of Tl and Pb in synthetic solutions containing both cations at concentration ratios from 1:5 to 5:1 at 10^{-6}–10^{-7} M concentration level. However, to the best of our knowledge, no application of these methods to complex natural samples (nonsynthetic) is available.

These interferences can be eliminated by addition of complexing agents and/or surfactants that form complexes with interfering ions but not with Tl(I). The more useful complexing agents are complexones (EDTA, DCTA, DTPA, DPTE, etc.), normally used together with acetates, citrates, tartrates, ascorbates, and so forth. The use of surfactants together with complexones yields additional selectivity because of the inhibition of other electrochemical processes.

Bonelli et al. (1980) reported a direct DPASV method, without previous steps, that can be used in the concentration ranges 0.5–100 μg Tl/L (HMDE) and 0.01–10 μg Tl/L (MFE). In both cases the measurements are made in EDTA–acetate buffer media. When the HMDE is used, the operating parameters are: conditioning at -0.1 V for 1.5 min with stirring; deposition at -0.8 V for 60 s with stirring, and equilibration at the same potential without stirring for 15 s. A well-defined DPASV peak at -0.42 V corresponds to the thallium oxidation, the height of which is proportional to the thallium concentration. For low thallium concentrations (0.01–10 μg/L) the authors recommend the use of MFE. Conditions are exactly as described above, except that electrode rotation at 3600 rpm is used instead of magnetic stirring, and the deposition time is increased to 600 s. In these conditions the DPASV peak appears at -0.56 V. The method has been used for routine thallium determinations in natural water samples and in several samples of atmospheric deposition (wet and dry precipitation) collected by the U.S. Geological Survey Central Laboratory of Denver. The same medium (acetate buffer and EDTA) has been used by Calderoni and Ferri (1982) to determine thallium in rocks with an HMDE. Liem et al. (1984) also determined thallium in rocks but in a medium of citrate and EDTA at pH 7–8. The determination of thallium in dust and rain in a medium of tartrate and EDTA at pH 4.5 with an HMDE has been described by Daneshwar and Zarapka (1980). Seitz et al. (1973) developed a method to determine thallium in seawater in a tubular mercury-covered graphite electrode by ASV in a flowing system. The determination is carried out in the presence of EDTA to mask other metals. The determination of Tl(I) in an HMDE using an acetate buffer pH 4.5 in the presence of EDTA has been accepted as an standard method in Germany (Deutsche Norm, 1988).

To increase selectivity, the simultaneous use of surfactant (action) and the complexing properties of the base electrolyte leads to the best results, like in polarographic methods. In this line You and Neeb (1983) determined Tl(I) in the presence of an excess of lead (about 10^4–10^5-fold) using as base electrolyte acetate buffer containing various complexones to complex lead, and several surfactants (Triton X-100, Eulan NK, Sulfetal). The method was checked with synthetic samples, obtaining in all cases linear calibration plots for Tl(I) in the concentration range from 10^{-6} to 10^{-8} M of Tl(I). Ciszewski and Lukaszewski (1985) described a method to avoid both the interferences of large quantities of copper and the simultaneous presence of copper, lead, and bismuth in the determination of Tl(I) by DPASV. The experiments were performed in an EDTA solution at pH 4.5. For the authors the most effective surfactants are ethoxyalkylphenols and ethoxyalcohols with different numbers of ethylene oxide units, poly(ethylene glycols) (PEG) and molecular weights above 2000 g (since PEGs with lower molecular weights are ineffective), as well as a group of ammonium salts and phosphonium salts having various degrees of symmetry. All the nonionic surfactants suppress the copper peak to much the same extent and shift it in the anodic direction. In contrast, the cationic surfactants exert widely differing suppressive effects.

Opydo (1989) used anionic surfactants in the determination of 10^{-8} M Tl(I) in the presence of 2–6 10^5-fold molar excesses of lead. The method was tested by determining the thallium content of soil extracts. As in other proposed methods, the supporting electrolyte was a pH 4.4 buffer solution containing EDTA. Among the different anionic surfactants studied, only Malester, Rokanol PL-10 and alkyl benzene sulfonate (ABS) caused damping in the signal of 5×10^{-3} M Pb(II) while the height of thallium remained unaffected. This makes the determination of 10^{-8} M Tl(I) possible when the molar excess of lead is about 5×10^5-fold. Other anionic surfactants gave satisfactory results, but they only damped 2×10^{-3} M of lead. The operating parameters for the DPASV determination are 10 min of electrodeposition in an HMDE at -0.7 V and a rest period of 20 s. The peak appears at -0.5 V.

The interferences caused by the presence of iron and/or titanium are difficult to overcome even though these metals do not accumulate in the mercury. These elements are present in many matrices at very high concentrations, which frequently prevents the determination of other elements. These interferences can be limited by the use of properly selected surfactants, but sometimes the simultaneous presence of several interferences is difficult to overcome; the use of EDTA, which limits the interference of lead, involves the appearance of the titanium peak, which does not appear in the noncomplexing media (Ciszewski and Lukaszewski, 1988). Lukaszewski and Zembrzuski (1992) developed a method that removes the interferences of iron, aluminum, and manganese by media exchange performed in a flow injection measuring system, and other interferences are removed by the use of a base electrolyte consisting of 0.15 M EDTA and 0.1 M ascorbic acid. The operating procedure consists of the transfer of 8 mL of sample solution in the flow injection system. The solution was directed to the measuring cell and then back to the beaker with the sample. In this way the solution was in circulation. The preconcentration was started by application of a potential of -0.85 V to MFE for 1–3 min, depending on the thallium concentration. After the deposition step the analyzed solution was replaced by the deaerated carrier solution (0.05 M EDTA). Then the flow was stopped and the voltammogram was recorded after 15 s. This method was applied successfully to the determination of thallium in soils, as an example of complicated environmental matrices with large content of possible interferences. The content of Tl(I) in the tested samples was between 100 and 350 ppb.

Von Laar et al. (1994) used an MFE (plated ex situ on a glassy carbon surface) to choose a deposition potential close to the half-wave potential of Tl(I) that eliminated the interferences of lead, cadmium, and iron in the determination of Tl(I) in soils, using a media containing EDTA and ascorbic acid. The difference with the previous method by Lukaszewski and Zembrzuski (1992) is that here the plating procedure of the mercury film electrode and the accumulation of Tl(I) and other metals are carried out (subsequently) in two steps, whereas in the previous method a medium exchange after the deposition step is included to reduce interferences.

The application of these methods to natural samples not only requires the consideration of interferences from (coexisting) metal ions and organic compounds in the samples, but must also consider the low concentrations (nanogram per gram) in such samples. That is why some methods combine the final measurement with preconcentration procedures. Moreover, some authors propose the use of MFE to increase sensitivity, in spite of its lower reproducibility with respect to HMDE.

Batley and Florence (1975) determined thallium in seawater and nonsaline river water by DPASV in an HMDE using an acetate buffer pH 4.6–EDTA medium, after a cleanup with an anion exchange resin. The method requires 3 days of preconcentration and a volume of sample of 4 L to obtain a detection limit of 0.6 ng/L. After 10 mins of preelectrolysis at -0.8 V (vs. SCE) a DPASV peak at -0.45 V is obtained. Shevchenko et al. (1977) proposed a method that involves the preconcentration of thallium by coprecipitation with $Mg(OH)_2$, followed by ASV at MFE. The method requires 3 h of preconcentration, and 100 mL of sample, but the detection limit is 10 ng/L.

Although thallium is usually present as Tl(I) in aqueous solutions, Tl(III) has to be considered in some cases. Hoeflich et al. (1983) determined simultaneously Tl^+ and $(CH_3)_2 Tl^+$ in laboratory samples by DPP and DPASV. Barisci and Wallace (1992) proposed a photoelectrochemical method for the speciation and electrochemical detection of Tl(I) and Tl(III) in flowing solutions [which is not always possible with the usual ASV methods due to the preliminary reduction of all thallium species to $Tl^0(Hg)$]. For this purpose, the authors took advantage of a photochemical reaction between Tl(III) and some carboxylic acids like formic, oxalic, and glycoxylic, that had been previously applied to the determination of Tl(I) and Tl(III) by titrimetric analysis (Sagi et al., 1982).

The method is based on (i) the oxidation of Tl(I) to Tl(III) on a Pt electrode at 1.1 V (vs. Ag/AgCl), and (ii) the reduction of Tl(III) to Tl(I) with 0.05 M formic acid in a strong acidic media (0.25 M H_2SO_4), which only takes place at a significant rate in the presence of UV radiation. These reactions can be written as:

$$(i)\ Tl^+ \rightarrow Tl^{3+} + 2e^-$$

$$(ii)\ Tl^{3+} + HCOOH + h\nu \rightarrow Tl^+ + CO_2 + 2H^+$$

According to this scheme, when a positive potential is applied to the platinum electrode in the absence of ultraviolet (UV) radiation only Tl(I) ions are oxidized and thus detected. When the same process is carried out under UV radiation, Tl(III) ions are photochemically converted into Tl(I) in the flowing solution, so that they also contribute to the oxidation signal on the electrode. Then, discrimination between Tl(I) and Tl(III) is possible by detection with and without UV irradiation.

Under such conditions, calibration plots for both Tl(I) and Tl(III) were linear up to 10 ppm, and detection limits were of the order of 10 ppb for both ions.

As mentioned above, functional groups at the surface of the working electrode may increase the selectivity and the sensitivity of common voltammetric methods. In particular, the possibility to preconcentrate electroactive species due to chemical or physico-chemical reactions with the modifier often allows very small amounts of analyte to be detected. Additionally, in some cases, such accumulations do not require any applied potential and, may, therefore, be performed under open-circuit conditions including a medium exchange step preceding the final measurement.

Cai and Khoo (1995) prepared a carbon paste electrode modified with 8-hydroxyquinoline to selectively accumulate Tl(I) in open circuit. The accumulation of thallium can be viewed as a process of complexation/adsorption at the electrode surface. During the reduction step, the surface bound Tl(I) was reduced to metallic thallium, which was subsequently reoxidized to Tl(I) during the stripping step. Under optimum conditions (differential pulse scan in the stripping step and 120 s of accumulation) a detection limit of 4.9 $\times$ 10^{-9} M (1.00 ppb) was found. Interferences from Bi(III), Pb(II), and Cu(II) were removed by masking with complexing agents such as sodium diethyldithiocarbamate, EDTA, and sodium citrate. The authors also proposed the use of this method for thallium speciation. The modified method is based on the determination of Tl(I) while Tl(III) is masked with EDTA, followed by chemical reduction of Tl(III) to Tl(I) with hydroxylamine hydrochloride and further determination of total thallium.

A method was proposed by Diewald et al. (1993) to determine thallium as tetrachlorothallate(III) by preconcentration from 10^{-3} M HCl containing Tl(III) onto a carbon paste electrode that had been chemically modified with an anion exchanger (Amberlite LA-2). The accumulation was performed under open-circuit conditions. After accumulation, thallate(III) ions are reduced to Tl(0) at -1 V (vs. SCE). The element is reoxidized to Tl(I) during the scan in the anodic direction, yielding a sharp signal response at -0.86 V.

6.4.3. Adsorptive Stripping Voltammetry

Wang and Lu (1993) determined thallium by AdSV using o-cresolphthalein (OCP) to complex and adsorb Tl(I) onto the electrode. Optimal conditions were found to be 0.01 M sodium hydroxide containing $5 \times 10^{-7} M$ OCP, a deposition potential of -0.3 V and a differential pulse sweep until -0.9 V in the stripping step. In these conditions, a detection limit of 60 ng/L was found for a deposition time of 2 mins. The method proved not to be sensitive to the presence of Pb(II), In(III), Sn(IV), Ni(II), Zn(II), Cr(VI), Se(IV), Al(III), Fe(III), Te(IV), Mo(VI), Mg(II), U(VI), and V(V). However, surfactants can markedly decrease the stripping peak.

6.4.4. Abrasive Stripping Voltammetry

Zhang et al. (1995) used AbrSV to identify different species of thallium and other metals in solid samples. For this purpose, the powdered samples were rubbed with a graphite electrode, the pores of which had been closed and filled with paraffin (paraffin-impregnated graphite electrode, PIGE). After the sample transfer to the electrode surface, the PIGE was introduced into an electrochemical cell containing a 2 M NaOH solution, and a potential sweep toward the negative direction was applied in order to reduce thallium species. The main drawback of the technique is the lack of strict control of the amount of sample transferred to the electrode. However, it can be useful for qualitative purposes. A more powerful tool is abrasive coulometry (i.e., the measurement of the charge required to oxidize or reduce the deposited sample), which can provide the stoichiometry of some thallium compounds, such as sulfides of thallium or thallium–tin sulfosalts.

6.5. POTENTIOMETRIC STRIPPING ANALYSIS

6.5.1. Introduction

Potentiometric stripping analysis (PSA) was reported by Jagner and Granelli (1976) as a novel analytical technique for the determination of metal traces.

Although this technique should, more accurately, be referred as chronopoten-tiometric stripping analysis, it is usually called PSA.

Potentiometric stripping analysis is a two-step technique consisting of an electrolysis step and a stripping step, and it is closely related to the previously described ASV. In the electrolysis step metal ions are reduced to free metal and deposited as amalgam at the working electrode, commonly an MFE. As in ASV, the electrolysis process is primarily controlled by the concentration and the diffusion coefficient of the metal ions involved, and the thickness of the diffusion layer at the working electrode as imposed by the stirring regime. In contrast to ASV, the stripping step is essentially a chemical process in which the electric circuit is interrupted and a chemical oxidant, such as oxygen or Hg(II) ions, will reoxidize the amalgam to metal ions. In such a process, the change in the electrode potential E with time is monitored. Integration of the dt/dE function between two plateau values of the potential yields the total stripping time t_s for the metal investigated. Figure 9 and Table 1 summarize the main features in PSA.

Modern computerized PSA instruments allow measurement of the potential up to 30,000 times per second.

The phenomena involved in the PSA cycle at an MFE have been studied theoretically (Jagner, 1982; Chau et al., 1982; Labar, 1993) in order to derive equations for the E–t relationship. These rigorous treatments are beyond the scope of this chapter, but, briefly, the stripping time, t_s, can be expressed as:

$$t_s = k[\mathrm{Me}^{n+}]t_D d_s[\mathrm{Ox}]^{-1} d_D^{-1},$$

where k is a constant including a number of physical constants such as: (i) the diffusion coefficients of the analyte Me^{n+} and the oxidizing agent Ox;

Potential vs. time

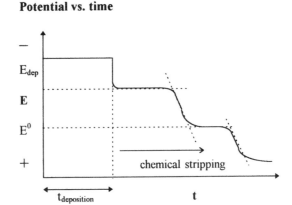

Figure 9. Sequence for potentiometric stripping analysis.

(ii) the hydrodynamic constants depending on the electrode and cell geometries and the stirring regime during the different steps in the analysis; and (iii) the ratio of the number of electrons involved in the reduction and oxidation of the metal Me. [Me^{n+}] denotes the bulk concentration of the analyte, t_D is the deposition time, [Ox] is the concentration of the oxidant, and d_D and d_s are the diffusion layer thicknesses during deposition and stripping, respectively.

For main analytical purposes, there is a domain of experimental values of [Ox], t_D, d_D, and d_s, in which t_s and [Me^{n+}] are linearly related.

The basic experimental setup needed for a PSA experiment comprises a potentiostatic circuit, a three-electrode electrochemical cell, and an impedance recorder. However, to improve performances, commercially available instruments have been introduced as well as customized microcomputer-controlled configurations built from standard electrochemical equipment. The potential of the working electrode is usually sampled at frequencies from 7 to 26 kHz.

Several variants of PSA technique have been developed, whose details and related theoretical aspects can be found elsewhere (Estela et al., 1995).

Although PSA is not a very popular analytical technique for the determination of metal traces, in samples containing high concentrations of inorganic salts, PSA can be a very valuable technique. Typical samples are body fluids, mineral acid digests from biological materials, saline waters, and chemicals for purity testing. Of special interest in this context are the reviews by Jagner (1982) and Estela et al. (1995), as well as recent work of Ostapczuk (1993) and Labar (1993).

Applications of PSA are mainly devoted to the more ubiquous trace metals such as Cu, Pb, Cd, and Zn. Applications to other trace metals such as Tl, Bi, Sn, Sb, In, and the like are much more scarce. Concerning the determination of Tl, some previous investigations can be mentioned as representative of the work made at the moment.

6.5.2. Applications

Christensen et al. (1982) showed the suitability of PSA for the determination of Cd, Pb, and Tl in fly ash and in waters polluted by leaching of fly ash. The accuracy of the method developed was tested by analyzing a certified coal fly ash (NBS SRM 1633a).

Later, Jaya et al. (1985) developed a PSA procedure for the simultaneous determination of Pb and Tl in 0.1 M chloroacetic acid–0.5 M sodium chloride medium that gave the maximum signal for Pb and Tl with sufficient resolution. The method was tested for various mixtures of Pb and Tl in ratios ranging from 100:1 to 1:100. Seawater samples collected from the Bay of Bengal were analyzed. No values were reported for these seawater samples because the results obtained for spiked seawater samples gave the concentration values corresponding to the spiked amounts. The detection limit for Tl was not referred. From calibration graphs, the coefficient of variation for 10 replicate determination of 0.1 mg/L of Tl was 1.2%. The authors claimed that the method is suitable for seawater analysis.

Ostapczuk (1993) demonstrated that for PSA determination of trace elements in natural water samples, simple acidification with HCl is sufficient to liberate trace elements from the complexing agents. Thus, it is possible to determine Tl in natural waters at a pH of ca. 4.5 after the addition of EDTA to mask the Pb. The detection limit at 30 min deposition time was 0.1 ng/g, using an MFE. This is the order of magnitude of the values determined in natural water samples collected in the Environmental Specimen Bank at Jülich (Germany). In routine work in such a center, Tl can be determined by PSA in solid samples after digestion in polytetrafluoroethylene (PTFE) Teflon bombs or microwave digestion.

More recently, Cleven and Fokkert (1994) optimized the PSA experimental parameters for the determination of Tl in Rhine water sampled at Lobith at the Dutch/German border. Preliminary studies showed that the Tl concentration in Rhine water samples was lower than the operational detection limits. Therefore the samples were concentrated 10-fold by evaporation. The resulting average concentration in the original Rhine water sample was calculated to be 0.35 nmol/L, and the relative standard deviation σ after repetitive plating/stripping 10 measurements appeared to be 22%. Under the conditions of the standard settings adopted in this study, the 3σ value for repeated measurements of a low concentration of the standard solution gives an estimate of the operational detection limit for Tl of about 2×10^{-9} mol/L.

In other cases, attention was focused in the development of instrumental systems, then Tl was used as an analyte to test the reliability of the developed systems. Thus, for instance, Anderson et al. (1982) developed a thin-layer cell with a Hg-coated glassy carbon electrode that allowed the determination, in synthetic samples, of Tl in a 10^4-fold excess Cd. Later, Hoyer et al. (1986) developed a computerized system with a hard-wired data acquisition unit for PSA. The utility of the instrument was demonstrated by analyzing the certified fly ash (NBS SRM 1633a), which has a low Tl content but a very high content of oxidants [e.g., 9.4% of Fe(III)], which causes rapid stripping of the amalgamated analytes.

6.6. SUMMARY

A brief overview has been given of some of the research that has been made, and is currently taking place, in the field of thallium determination by electroanalytical techniques. In recent years great efforts have been devoted to the application of AdSV, AbrSV, PSA, and CMEs. It seems reasonable to expect that improvements in such subjects, especially by incorporating more complicated molecular ensembles and by micro- and nano-preparation, will render benefits for the determination of the very low total thallium concentration and also for its speciation.

ACKNOWLEDGMENTS

The authors gratefully acknowledge financial support from the Ministerio de Educación y Cultura of Spain (DGICYT, Project PB93–1055) and from the Direcció General de Universitats de la Generalitat de Catalunya (GRQ 1028).

REFERENCES

Anderson, L., Jagner, D., and Josefson, M. (1982). Potentiometric stripping analysis in flow cells. *Anal. Chem.*, **54**, 1371–1376.

Arrigan, D. W. M. (1994). Voltammetric determination of trace metals and organics after accumulation at modified electrodes. *Analyst*, **119**, 1953–1966.

Barendrecht, E. (1967). Stripping voltammetry. In *Electroanalytical Chemistry*, Vol. 2, Bard, A. J. (ed), Marcel Dekker, New York, pp. 53–109.

Barisci, J. N., and Wallace, G. G. (1992). Photoelectrochemical detection and speciation of thallium (I) and thallium (III). *Electroanalysis*, **4**, 139–142.

Batley, G. E., and Florence, T. M. (1974). Evaluation and comparison of some techniques of anodic stripping voltammetry. *J. Electroanal. Chem.*, **55**, 23–43.

Batley, G. E., and Florence, T. M. (1975). Determination of thallium in natural waters by anodic stripping voltammetry. *J. Electroanal. Chem.*, **61**, 205–211.

Bond, A. M. (1980). *Modern Polarographic Methods in Analytical Chemistry.* Marcel Dekker, New York.

Bonelli, J. E., Taylor, H. E., and Skogerboe, R. K. (1980). A direct differential pulse anodic stripping voltammetric method for the determination of thallium in natural waters. *Anal. Chim. Acta*, **118**, 243–256.

Cai, Q., and Khoo, S. B. (1995a). Determination of trace thallium after accumulation of thallium (III) at a 8-hydroxyquinoline-modified carbon paste electrode. *Analyst*, **120**, 1047–1053.

Cai, Q., and Khoo, S. B. (1995b). Differential pulse stripping voltammetric determination of thallium with an 8-hydroxyquinoline-modified carbon paste electrode. *Electroanalysis*, **7**, 379–385.

Calderoni, G., and Ferri, T. (1982). Determination of thallium at subtrace level in rocks and minerals by coupling differential pulse anodic-stripping voltammetry with suitable enrichment methods. *Talanta*, **29**, 371–375.

Chau, T. C., Li, D., and Wu, Y. L. (1982). Studies on potentiometric stripping analysis. *Talanta*, **29**, 1083–1087.

Christensen, J. K., Kryger, L., and Pind, N. (1982). The determination of traces of cadmium, lead and thallium in fly ash by potentiometric stripping analysis. *Anal. Chim.Acta*, **141**, 131–146.

Ciszewski, A, and Lukaszewski, Z. (1985). Electrochemical masking of large amounts of copper in DPASV and the determination of thallium in the presence of large excess of copper. *Talanta*, **32**, 1101–1104.

Ciszewski, A., and Lukaszewski, Z. (1988). Electrochemical masking for removal of the titanium matrix effect in DPASV determination of thallium. *Talanta*, **35**, 191–197.

Cleven, R., and Fokkert, L. (1994). Potentiometric stripping analysis of thallium in natural waters. *Anal. Chim. Acta*, **289**, 215–221.

Coetzee, C. J. (1985). Properties and analytical application of a thallium(I) ion-selective electrode. *Talanta*, **32**, 821–823.

Coetzee, C. J., and Basson, A. J. (1973). Potentiometric studies with thallium(I)–heteropoly acid salt–epoxy resin membranes. *Anal. Chim. Acta*, **64**, 300–304.

Coetzee, C. J., and Basson, A. J. (1977). Cesium and thallium(I)-sensitive liquid membrane electrodes based on cesium and thallium tetrakis(*m*-trifluoromethylphenyl)borates. *Anal. Chim. Acta,* **92,** 399–403.

Cotton, F. A., and Wilkinson, G. (1966). *Advanced Inorganic Chemistry,* 2nd ed. Wiley, New York.

Dan, D., and Dong, Y. (1988). A PVC-coated carbon rod ion-selective electrode for thallium and its application to the analysis of rocks and minerals. *Talanta,* **35,** 589–590.

Daneshwar, R. G., and Zarapka, L. P. (1980). Simultaneous determination of thallium and lead at trace levels by anodic-stripping voltammetry. *Analyst,* **105,** 386–390.

Deutsche Norm (1988). DIN 38 406, Teil 6, March.

Dieker, J., and van der Linden, W. E. (1975). Determination of Tl(I), Pb(II) and Cu(II) by differencial pulse anodic stripping voltammetry with a thin mercury film electrode. *Fresenius Z. Anal. Chem.,* **274,** 97–101.

Diewald, W., Kalcher, K., Menhold, C., Cai, X., and Magee, R. J. (1993). Voltammetric behaviour of thallium(III) on carbon paste electrodes chemically modified with an anion exchanger. *Anal. Chim. Acta,* **273,** 237–244.

Eisenman, G. (1965). The electrochemistry of cation-sensitive glass electrodes. In *Advances in Analytical Chemistry and Instrumentation,* Vol. 4, Reilley, C. N. (ed). Wiley, New York.

Estela, J. M., Tomas, C., Cladera, A., and Cerdà, V. (1995). Potentiometric stripping analysis: A review. *Crit. Rev. Anal. Chem.,* **25,** 91–141.

Fogg, A. G., Al-Sibaai, A. A., and Burgess, C. (1975). Ion-selective electrodes for the determination of antimony and thallium based on water-insoluble basic dye salts. *Anal. Lett.,* **8,** 129–137.

Fombon, J. J., Oddon, Y., Tacussel, J., and Tranquard, A. (1979). Étude et mise au point de deux types d'électrodes spécifiques à élément sensible polycristallin pour la détermination des ions thallium(I). *Analusis,* **7,** 494–498.

Freiser, H. (1978). *Ion-selective Electrodes in Analytical Chemistry.* Plenum, New York.

Freiser, H., and Nancollas, G. H. (eds.) (1987). *IUPAC Compendium of Analytical Nomenclature. Definitive Rules 1987.* Blackwell, Oxford.

Grabarić, Z., Grabarić, B. S., Esteban, M., and Casassas, E. (1995). Signals ratio method for resolution enhancement in differential pulse polarography and related techniques. *Anal. Chim. Acta,* **312,** 27–34.

Henrion, A., Henrion, R., Henrion, G., and Scholz, F. (1990). Application of partial least-squares regression for signal resolution in differential pulse anodic stripping voltammetry of thallium and lead. *Electroanalysis,* **2,** 309–312.

Hoeflich, L. K., Gale, R. J., and Good, M. L. (1983). Differential pulse polarography and differential pulse anodic stripping voltammetry for determination of trace levels of thallium. *Anal. Chem.,* **55,** 1591–1595.

Hoyer, B., Skov, H. J., and Kryger, L. (1986). Computerized system with a hard-wired data-acquisition unit for potentiometric stripping analysis. *Anal. Chim. Acta,* **188,** 205–217.

Huang, D., Zhu, C., Zhang, J., Lei, H., Wang, D., Hu, H., Fu, T., Ou, H., Shen, Z., and Yu, Z. (1984). Preparation of bis(crown ether) PVC membrane thallium(I) ion-selective electrodes. *Fenxi Huaxue,* **12,** 89–93; *Chem. Abs.,* **100,** 202566h.

Jacobsen, E., and Kalland, G. (1963). Polarographic determination of thallium. *Anal. Chim. Acta,* **29,** 215–219.

Jagner, D. (1982). Potentiometric stripping analysis. A review. *Analyst,* **107,** 593–599.

Jagner, D., and Graneli, A. (1976). Potentiometric stripping analysis. *Anal. Chim. Acta,* **83,** 19–26.

Jain, A. K., Singh, R. P., and Agrawal, S. (1980). Membranes of α-picolinium molybdoarsenate in Araldite as a thallium(I) ion-selective electrode. *Fresenius Z. Anal. Chem.,* **302,** 407–409.

Jaya, S., Prasada Rao, T., and Prabhakara Rao, G. (1985). Simultaneous determination of lead and thallium by potentiometric stripping. *Talanta,* **32,** 1061–1063.

Khalil, M. M., Tănase, I., and Luca, C. (1985). A polarographic study of some complexes of Tl(I) with polyoxa macrocyclic ligands. *Talanta*, **32**, 1151–1152.

Khopkar, S. M., and Gandhi, M. (1995). Thallium. In *Encyclopedia of Analytical Science* Vol. 9, Townshend, A. (ed). Academic Press, London, pp. 5132–5135.

Klahre, P., Valenta, P., and Nürnberg, H. W. (1978). Ein normiertes pulsinversvoltammetrishes analysenverfahren zur prüfung von trinkwasser auf toxische metalle. I. Simultanbestimmung von kupfer, cadmium, blei und zink und von blei und thallium. *Vom Wasser*, **51**, 199–219.

Koryta, J. (1972). Theory and applications of ion-selective electrodes. *Anal. Chim. Acta*, **61**, 329–411.

Koryta, J. (1977). Theory and applications of ion-selective electrodes. Part II *Anal. Chim, Acta*, **91**, 1–85.

Koryta, J. (1979). Theory and applications of ion-selective electrodes. Part III *Anal. Chim. Acta*, **111**, 1–56.

Koryta, J. (1982). Theory and applications of ion-selective electrodes. Part 4ª. *Anal. Chim. Acta*, **139**, 1–51.

Koryta, J. (1984). Theory and applications of ion-selective electrodes. Part 5ª. *Anal. Chim. Acta*, **159**, 1–46.

Koryta, J. (1986). Theory and applications of ion-selective electrodes. Part 6ª. *Anal. Chim. Acta*, **183**, 1–46.

Koryta, J. (1988). Theory and applications of ion-selective electrodes. Part 7ª. *Anal. Chim. Acta*, **206**, 1–48.

Koryta, J. (1990). Theory and applications of ion-selective electrodes. Part 8ª. *Anal. Chim. Acta*, **233**, 1–30.

Koryta, J., and Mittal, M. L. (1972). Electroreduction of monovalent metal ion complexes of macrocyclic polyethers. *J. Electroanal. Chem.*, **36**, App. 14–19.

Labar, C. H. (1993). Potentiometric stripping analysis and speciation of heavy metals in environmental studies. *Electrochim. Acta*, **38**, 807–813.

Liem, I., Kaiser, G., Sager, M., and Tölg, G. (1984). The determination of thallium in rocks and biological materials at ng g^{-1} levels by differential-pulse anodic stripping voltammetry and electrothermal atomic absorption spectrometry. *Anal. Chim. Acta*, **158**, 179–197.

Lukaszewski, Z., and Zembrzuski, W. (1992). Determination of thallium in soils by flow-injection-differential pulse anodic stripping voltammetry. *Talanta*, **39**, 221–227.

Malik, W. U., Srivastava, S. K., Razdan, P., and Kumar, S. (1976). Tungstoarsenates as ion-selective membranes for cesium and thallium(I) ions. *J. Electroanal. Chem.*, **72**, 111–116.

Masuda, Y., Yakabe, K., Shibutani, Y., and Shono, T. (1994). Thallium(I) ion-selective electrode based on polythiamacrocycles. *Anal. Sci.*, **10**, 491–495.

Murray, R. W. (1984). Chemically modified electrodes. In *Electroanalytical Chemistry*, Vol. 13, Bard, A. J. (ed). Marcel Dekker, New York, pp. 191–368.

Neeb, R., and Kiehnast, I. (1968). Das invers-voltammetrische verhalten einiger elemente in salzsauren grundlösungn. Anodische amalgamvoltammetrie VIII. *Fresenius Z. Anal. Chem.*, **241**, 142–155.

Opydo, J. (1989). Determination of thallium in the presence of large excess of lead by anodic stripping voltammetry. *Mikrochim. Acta*, **II**, 15–22.

Ostapczuk, P. (1993). Present potentials and limitations in the determination of trace elements by potentiometric stripping analysis. *Anal. Chim. Acta*, **273**, 35–40.

Osteryoung, J. G., and Osteryoung, R. A. (1985). Square wave voltammetry. *Anal. Chem*, **57**, 101A–110A.

Přibil, R., and Zábranský, Z., (1951). Polarographic determination of thallium. *Coll. Czech. Chem. Commun.*, **16**, 427–430.

Paneli, M. G., and Voulgaropoulos, A. (1993). Applications of adsorptive stripping voltammetry in the determination of trace and ultratrace metals. *Electroanalysis*, **5**, 355–373.

Parham, H., and Shamsipur, M. (1991). Polarographic study of the interaction between heavy metal ions and some macrocyclic ligands in binary acetonitrile + water mixtures. *J. Electroanal. Chem.*, **314**, 71–80.

Pižeta, I., Jeren, B., and Aleksić-Maslać, K. (1994). Straight lines, windows and background current synthesis in deconvolution porcedure. *J. Electroanal. Chem.*, **375**, 169–174.

Pižeta, Y. (1994). Deconvolution of non-resolved voltammetric signals. *Anal. Chim. Acta*, **285**, 95–102.

Reilley, C. N., Scribner, W. G., and Temple, C. (1956). Amperometric titration of two- and three-component mixtures of metal ions with (ethylenedinitrilo)tetraacetic acid. *Anal. Chem.*, **28**, 450–454.

Sagi, S. R., Prakasa Raju, G. S., Appa Rao, K., and Prasada Rao, M. S. (1982). A photochemical redox method for the estimation of thallium(III). *Talanta*, **29**, 413–415.

Scholz, F., and Lange, B. (1992). Abrasive stripping voltammetry- an electrochemical solid state spectroscopy of wide applicability. *Trends Anal. Chem.*, **11**, 359–367.

Seitz, W. R., Jones, R., Klatt, L. N., and Mason, W. D. (1973). Anodic stripping voltammetry at a tubular mercury-covered graphite electrode. *Anal. Chem.*, **45**, 840–844.

Shetty, P. S., Subbaraman, P. R., and Gupta, J. (1962). The polarographic determination of thallium. *Anal. Chim. Acta*, **27**, 429–433.

Shevchenko, L. V., Portretnyi, V. P., and Chuiko, V. T. (1977). Determination of thallium in natural water by inverse voltammetry with preconcentration by coprecipitation. *J. Anal. Chem. USSR*, **12**, 1448–1450; *Chem Abs.*, **87**, 189134g.

Szczepaniak, W., and Ren, K. (1980). Analytical application of a thallium(I)-selective membrane electrode. *Chem. Anal. (Warsaw)*, **25**, 449–458; *Chem. Abs.*, **94**, 40815s.

Szczepaniak, W., and Ren K. (1976). A thallium(I)-selective electrode based on a liquid ion-exchanger containing *o,o'*-didecyldithiophosphoric acid. *Anal. Chim. Acta*, **82**, 37–44.

Tamura, H., Kimura, K., and Shono, T. (1980). Thallium(I)-selective PVC membrane electrodes based on bis(crown ether)s. *J. Electroanal. Chem.*, **115**, 115–121.

Vesely, J., Weiss, D., and Stulik, K. (1978). *Analysis with Ion-Selective Electrodes*. Ellis Horwood, Chichester.

Vlasov, Yu. G., Kolodnikov, V. V., and Chernykh, V. G. (1981). Ion-selective electrodes based on mercury(II) and thallium iodides for determining thallium(I) ions. *J. Anal. Chem. USSR*, **36**, 1319–1322.

von Laar, C., Reinke, R., and Simon, J. (1994). Determination of thallium in soils by differential pulse anodic stripping voltammetry by means of a mercury film electrode. *Fresenius J. Anal. Chem.*, **349**, 692–693.

Vydra, F., Stulik, K., and Jukálová, E. (1976). *Electrochemical Stripping Analysis*. Ellis Horwood, Chichester.

Wang, J. (1985). *Stripping Analysis. Principles, Instrumentation and Applications*. VCH, Deerfield Beach, FL.

Wang, J., and Lu, J. (1993). Adsorptive stripping voltammetry of trace thallium. *Anal. Chim. Acta*, **282**, 329–333.

Whitfield, M., and Turner, D. R. (1981). Sea water as an electrochemical medium. In *Marine Electrochemistry*, Whitfield, M., and Jagner, D. (Eds.). Wiley, New York.

Wilgocki, M., and Cyfert, M. (1989). Polarographic determination of lead(II) in the presence of thallium(I) and cadmium(II) using complexation with ethanediamine and hydroxyl ion. *Anal. Chim. Acta*, **226**, 351–358.

Yang, S., Tong, H., and Liu, P. (1987). Crown ether PVC membrane thallium(I) ion-selective electrodes. *Fenxi Huaxue*, **15**, 659–662; *Chem. Abs.*, **108**, 86968e.

You, N., and Neeb, R. (1983). Verbesserung der invers-voltammetrischen bestimmung des thalliums neben blei durch electrokhemische maskierung. *Fresenius Z. Anal. Chem.,* **314,** 394–397.

Zhang, S., Meyer, B., Moh, G. H., and Scholz, F. (1995). Development of analytical procedures based on abrasive stripping coulometry and voltammetry for solid state phase microanalysis of natural and sinthetic tin-, arsenic-, and antimony-bearing sulfosalts and sulfides of thallium, tin, lead, and silver. *Electroanalysis,* **7,** 319–328.

Zirino, A. (1981). Voltammetry of natural sea water. In *Marine Electrochemistry,* Whitfield, M., and Jagner, D. (Eds). Wiley, New York.

7

DETERMINATION OF THALLIUM AND ITS SPECIES IN AQUATIC ENVIRONMENTAL SAMPLES

Chiu L. Chou and John D. Moffatt

Marine Environmental Sciences Division, Science Branch,
Maritimes Region, Fisheries and Oceans Canada, P.O. Box 550,
Halifax, Nova Scotia, Canada, B3J 2S7

Thallium in the Environment, Edited by Jerome O. Nriagu.
ISBN 0-471-17755-5. © 1998 John Wiley & Sons, Inc.

7.1. INTRODUCTION

The present review briefly describes the available literature and database of thallium in the aquatic environment. Analytical techniques used for the determination of thallium in aquatic environmental samples are also discussed.

The method for differentiating Tl(I) and Tl(III) in aquatic samples is of considerable importance due to the high toxicity of thallium. Analytical procedures for thallium speciation are also reported in this chapter.

7.2.   LITERATURE OF THALLIUM DATA

Thousands of investigations concerning toxic metal concentrations in aquatic plants, living organisms, humans, soil, air, water, and wastewater in the environment have been published. Most of these publications are on toxic substances such as the transition metals (Cu, Zn, Cd, Hg, Pb, Ni, Cr, etc.) but very few address thallium pollution. The literature is almost devoid of data on thallium concentrations in the pelagic environment (Flegal et al., 1986). In a review of elemental distributions in sea water (Quinby-Hunt and Turekian, 1983), thallium was to be the only element without a published value. Also, there are very limited thallium concentration data for marine organisms (Zitko and Carson, 1975; Hamaguchi et al., 1960; Chou and Uthe, 1995), marine teleostei (Heit, 1979), and marine algae and marine phytoplankton (Flegal et al., 1986).

The lack of studies on thallium concentrations in the marine environment may be attributed to:

1. The relatively small range and extremely low concentrations in sea water (12–16 ng/L)
2. The limitation (cost and availability) of ultrasensitive instrumentation for detection [such as inductively coupled plasma mass spectrometry (ICP-MS), isotope dilution mass spectrometry (ID-MS), and laser-excited atomic fluorescence spectrometry (LEAFS)]
3. The low bioaccumulation in marine living species.

In the study of thallium in marine plankton, Flegal et al. (1986) reported that the bioaccumulation of thallium in plankton relative to seawater is not noticeably reflected in its oceanic cycle. Also reported was that thallium has a short biological half-life in plankton, similar to that (3–8 days) in human beings (Wallwork-Barber et al., 1985).

7.3.   DISTRIBUTION AND OCCURRENCE IN THE NATURAL ENVIRONMENT

There are a number of review papers (Sager, 1992a, 1994) and a book (Smith and Carson, 1977) on the distribution and occurrence of thallium in the natural environment. Thallium is widely distributed in the environment in a large number of minerals and is usually found in sulfur-containing ores and potassium minerals (Merian, 1991; Chapter 1). Natural environmental levels of thallium are comparatively low. The mean concentration in soil is 0.25 μg/g (Ure and Berrow, 1982).

Thallium has a significant pollution potential for aquatic systems. World recovery of thallium from wastes is 9–11 metric tons/yr, but, according to Kogan (1970), 44 metric tons per year are present in wastes from zinc production in Canada alone. Mining of other metals, such as potash and silicates, also release thallium to the environment, and coal-fired power plants emit thallium in fly ash (Davison et al., 1974). It is found in coal ranging in concentration from 0.05 to 10.0 ppm (Bowen, 1966). Thallium is mainly released into the environment by fossil fuel combustion and as a by-product of the roasting of raw materials used in cement production. It is also introduced into the environment as waste from the production of zinc, cadmium, and lead and by the combustion of coal.

Thallium is used in low-temperature thermometers, photoelectric cells, dye pigments, a component of corrosion-resistant alloys, and various types of "resistant" cements. Other industrial uses of thallium include the manufacture of alloys, electronic devices, and special glass (Chapter 1). Many thallium-containing catalysts have been patented for industrial organic reactions (Zitko, 1975b).

7.3.1. Sea Water and Fresh Water

The level of soluble thallium present in the sea is between 9 and 16 ng/L, which is similar to that in fresh water (Chapters 2 and 5). In seawater at pH 8.1, the oxygen content is sufficient to oxidize Tl(I) to Tl(III) due to formation of trivalent chloro-complex. Batley and Florence (1975) reported that 80% of the thallium in the Pacific Ocean was found to occur in trivalent form and only 20% occurs as monovalent and alkyl-thallium compounds.

Thallium concentrations coastal sea water have been reported to be in the range of 0.9–1.1 μg/L (Miyazaki and Tao, 1991). Zitko et al. (1975) found thallium concentrations of 1–80 μg/L in two rivers that serve as drains for mining industries in New Brunswick, Canada. Mathis and Kevern (1975) found thallium concentrations of 2.1–23.1 μg/L in Wintergreen Lake in Kalamazoo, Michigan. They attributed this thallium to coal fly ash contamination.

7.3.2. Aquatic Animals

There are very few studies or results on the concentrations of thallium in fish tissues. In general, the concentrations of thallium in fish are very low. Vitaliano and Zdanowicz (1992) reported the concentrations of thallium in flounder egg in the range of <0.015 to <0.023 μg/g (dry weight). In their study using standard reference material DOLT-1 (dogfish liver) Tl concentration was <1.14 μg/g (below detection limit). In a study of thallium concentration in fish (sunfish, shiner, bullhead, and suckers) from the seepage basins that received liquid effluents from the separations areas, Loehle and Paller (1990) reported that all samples were below the detection limit of <0.1 to <0.5 μg/g. In view of human health concerns, Hsieh et al. (1982) have analyzed toxic metals (Cd, Hg, and Tl) in vegetables and marine fishes. They reported 2.6 μg/g in fresh fish using an amperometric titration method.

A remarkably high thallium concentration of 248 μg/kg (dry weight) in the reference material Mussels II 1981 from the German Environmental Specimen Bank was reported by Waidmann et al. (1990). The mussels were collected in a harbor on Lake Constance. The area was polluted with thallium from sewage and industrial sources. Chou and Uthe (1995) found extremely high thallium concentrations in digestive glands from lobsters captured within Belledune Harbor, New Brunswick, Canada. The harbor was polluted by the fertilizer plant and the lead smelter. The lobsters from the most contaminated site within the harbor had a mean digestive gland Tl concentration of 194 μg/kg (wet weight). This concentration was 20 times higher than the lobsters from the control site. Thallium concentration in lobster digestive gland from the fertilizer plant site was 45.6 μg/kg (wet weight). Considering that these are wet weights, the thallium concentration in these lobster digestive glands is very high.

7.3.3. Aquatic Plants

Thallium concentrations in algaes from the German Environmental Specimen Bank ranging from 14.3 to 24.8 μg/kg (dry weight) were reported (Waidmann et al., 1990). Zitko et al. (1975) found thallium concentrations of 9.5–43.4 μg/g dry weight in algae samples from the South Tomogonops River and the South Little Lake, New Brunswick, Canada. Two moss samples from the South Tomogonops River contained thallium at remarkably high values of 125 and 162 μg/g (dry weight). The significance of these levels of thallium is not known since the concentration of thallium in plant materials is seldom reported (Bowen, 1966).

7.3.4. Sediment and Soil

Thallium concentration in sediments from the German rivers Elbe and Rhine ranged from 0.83 to 5.19 μg/g, while BCR CRM "River Sediment" was found to contain 0.54 μg/g (Waidmann et al., 1992). Mathis and Kevern (1975) reported a mean Tl concentration of 13.1 μg/g (range 2.1–23.1 μg/g) in sediments from a eutrophic lake. This concentration is unusually high for sediments compared to a mean concentration of 0.1 μg/g in soil (Bowen, 1966).

The natural abundance of thallium in cultivated soils has been reported by Nriagu (1984) as 0.01–0.50 μg/g. Naganuma and Okutani (1991) found thallium concentrations of 0.18–0.15 μg/g in river sediments from Ichikawa City, Japan. Analyses of 106 coal samples and 26 slag and fly ash samples (Sager, 1993) revealed concentrations averaging 0.51 and 2.19 μg/g. Waidmann et al. (1990,

1992) determined thallium values in standard reference materials of 0.003–0.96 μg/g and in German river sediments at 0.53–5.2 μg/g. Allus and Brereton (1992) found thallium levels in cement dust at 0.39–0.7 μg/g and in sediments at 0.850 μg/g. Concentrations in geological samples ranged from 0.3 to 3.9 μg/g (Hubert and Chao, 1985). In fly ash from coal-fired power plants, the thallium concentration increases from 5–45 μg/g in larger particles to 29–76 μg/g in airborne flyash (Davison et al., 1974).

7.4. BIOLOGICAL EFFECTS

Thallium is moderately toxic to plants and highly toxic to mammals (Bowen, 1966). It does not occur normally in animal tissues, but it becomes cumulative if taken into an organism (Browning, 1969). In studying the toxicity of heavy metals to the marine diatom *Ditylum brightwellii (west) Grunow,* Canterford and Canterford (1980) reported that the order of metal toxicity, based on the free metal concentration, was $Hg^{+2} > Ag^+ > Cu^{+2} > Pb^{+2} > Cd^{+2} > Zn^{+2} > Tl^+$. The EC_{50} criteria, the metal concentration required to reduce growth by 50% as compared to the control, was used. There was a reduction in the toxicity of copper, zinc, cadmium, and lead to *D. brightwellii* with increasing concentration of ethylenediaminetetraacetic acid (EDTA). However, the increased concentration of EDTA had no effect on the toxicity of mercury, silver, and thallium.

Thallium is as acutely toxic to fish (e.g., Atlantic salmon) as is copper, on a weight basis (Zitko et al., 1975). Water hardness, which strongly affects the toxicity of metals to fish, may not have much effect on the toxicity of thallium because of its low complexing ability. The acute toxicities of copper and thallium, and of zinc and thallium to fish, are not additive (Zitko et al., 1975). In studies on thallium toxicity to embryos and larvae of fathead minnows, LeBlanc and Dean (1984) reported that no fathead minnow embryos survived exposure to 720 μg/L, but embryos were not affected by exposure to thallium concentrations up to 200 μg/L. Larval survival was significantly reduced from exposure to thallium concentrations as low as 40 μg/L, and none survived at 350 μg/L. LeBlanc and Dean (1984) concluded that the maximum acceptable toxicant concentration for thallium, as thallium sulfate, was less than 40 μg/L.

In studies on the effects of external Tl^+ with K^+, Pic et al. (1979) reported that Tl^+ ions can substitute for K^+ with a 10-fold higher affinity in activating the sodium pump in mullet red cells and mullet gill microsomal ATPase. However, Tl cannot mimic the effects of K^+ in stimulating Na^+ and Cl^- efflux across the gill.

7.5. METHODS OF ANALYSIS

Thallium measurement in environmental samples has been reported by a number of reviewers (Smith and Carson, 1977; Zitko, 1975a; Korenman, 1960).

These reports were carried out before 1980, and they provide intensive coverage of analytical methods for thallium determination by gravimetry, titrimetry, electrochemical techniques (i.e., polarography, anodic or cathodic stripping voltammetry), spectrofluorimetry, flame atomic absorption spectrophotometry, emission spectrography, X-ray technique, neutron activation analysis, and mass spectroscopy. Smith and Carson (1977) focused on the analytical problems and the lack of reference materials for the analysis of thallium in biological fluids, tissue, and environmental samples. As analytical instrumentation continues to become more advanced in automation, improvement of reproducibility, and sensitivity for the detection of trace thallium in environmental samples is possible. The present review is to include the most recent analytical methods after 1980 (i.e., ICP-MS, IDMS, and LEAFS) and to emphasize the analytical procedures and precautions that should be followed during the development of new analytical methods for the analysis of thallium. Current analytical procedures are discussed separately for each sample type.

7.5.1. Water

7.5.1.1. Graphite Furnace Atomic Absorption (GFAA) Spectrophotometry

Welz et al. (1988) used GFAA to investigate the use of a palladium nitrate–magnesium nitrate modifier for determining metals in natural water samples (Table 1). The influence of a wide variety of concomitants and spectral interferences on analyte elements was studied. For thallium analysis, the authors found that addition of lithium was necessary when sodium chloride was present in the samples. Lithium helped reduce thallium loss as the volatile chloride during the preatomization stage. High atomization temperatures resulted in a 50% loss in sensitivity due to increased diffusional losses. In addition, at high atomization temperatures, the Pd modifier caused a spectral interference when continuum source background correction was used. NaCl and $ZnCl_2$ in samples decreased the Tl signal.

De Ruck et al. (1988) determined thallium in natural waters by electrothermal atomic absorption spectrophotometry after preconcentration. Cerium sulfate was used to oxidize thallium, which was retained as the tetrachlorothallate(III) ion on an anion exchange column. Thallium was eluted with ammonium sulfite, acidified, evaporated, made to volume and analyzed. Relative standard deviation (RSD) was 1.2% with a detection limit of 3.3 ng/L.

Fangruo and Xiandeng (1994) developed a preconcentration technique followed by GFAAS to determine sub-ppm levels of thallium in natural waters. They used a polyurethane foam column to adsorb thallium and eluted with EDTA and $(NH_4)_2SO_4$ as a matrix modifier. Detection limit was 0.01 μg/L. Charring with EDTA causes some buildup of carbon on the tube, but this caused no significant interference in the determination of thallium.

Riley and Siddiqui (1986) determined thallium in natural waters and sea water by first preconcentrating it by adsorption from oxidizing medium onto

Table 1 Summary of Thallium Analysis in Water

Sample	Range [Tl]	Detection Limit	Technique	Sample Treatment/Comments	Reference
1. Natural water (SRM NIST 1643b)	8.1 μg/L	0.8 μg/L	GFAAS	Use of palladium nitrate plus magnesium nitrate modifier. Lithium necessary to reduce loss of thallium as volatile chloride during preatomization.	Welz et al., 1988
2. Natural waters Tap Ground Rain Mineral	123.1–176.0 ng/L <3.3–15.2 ng/L 17.5 ng/L <3.3–23 ng/L	3.3 ng/L	Electrothermal AAS	Preconcentration. Cerium-sulfate used to acidify Tl. Tl was retained as the tetrachlorothallate (III) ion on an anion exchange column.	de Ruck et al., 1988
3. Natural waters Mineral River	<0.01–0.030 μg/L 0.064–0.155 μg/L	0.01 μg/L	GFAAS	Preconcentration with polyurethane foam column to absorb Tl. Elution with EDTA and $(NH_4)_2SO_4$ for matrix modification. Carbon buildup on tube but no interference.	Fangruo and Xiandeng, 1994
4. Natural waters	8–56 ppt (ng/L)	1 ppt (ng/L)	LIF-GF	Collected samples in acid-washed plastic bottles. No pretreatment or preconcentration. Little or no matrix interferences.	Axner et al., 1993
5. Natural water Lake Pore samples	<40 ng/L	0.03 ng/L (0.6 fg)	LEAFS	Collected samples were acidified to 0.2% HNO_3 for determination.	Cheam et al., 1996b
6. Ottawa River Tap SLRS-2	6 ng/L 4 ng/L 5 ng/L	50 μg/g	ICP-MS	Column preconcentrations and elutions to remove alkali and alkaline earth elements.	Gwendy et al., 1996a,b
7. Natural water/ wastewater	<2–75 μg/L	N/A	Potentiometric voltammetry	Tl^+ selective glass sensor. No significant interference by other metal ions.	Miloshova et al., 1994

8. Groundwater	280–400 ppt (ng/L)	2 ppt (ng/L)	ICP-MS	Preconcentration on resin, elution with HNO_3, evaporation to dryness, and dissolution in 1% HNO_3. Sample introduction by an ultrasonic nebulizer.	Stetzenbach et al., 1994
9. River Mine Hot spring Water	28–270 ng/mL 15–160 ng/mL 33 ng/mL	1.3 ng/mL	ICP-AES	Tl extracted with DIBK after complexation with APDC-HMAHMDC.	Miyazaki et al., 1992
10. Antarctic snow	0.07–2.8 pg/g 0.23–20 pg/g	N/A	HR-ICP-MS and LIAF	No preconcentration, nor matrix interference.	Baiocchi et al., 1994
11. Arctic snow	0.3–0.9 pg/g	N/A	LEAFS	Samples were obtained with a Teflon cover, thawed, and acidified to pH 1.6 using double sub-boiled distilled HNO_3, and stored before direct determination of Tl.	Cheam et al., 1996a
12. Seawater	14.8–17.5 ng/L 15.2–17.0 ng/L	1 µg/L	GFAAS and DPASV	Preconcentration by absorption from oxidizing medium (bromine and HNO_3) onto strongly basic anion exchange column. Elution with sulfur dioxide, evaporation, chloroacetic acid/sodium chloride as a supporting electrolyte.	Riley and Siddiqui, 1986
13. Seawater	10 ng/mL	N/A	Potentiometric stripping voltammetry	Applied to determine the partitioning of Tl concentrations between particulates and the liquid phase in seawater.	Jaya et al., 1985
14. Seawater	12.08 ppt (ng/L)	N/A	IDMS	Direct injection into graphite furnace with palladium as matrix modifier to overcome halide interference.	Murozumi et al., 1982
15. Wastewater	1.4–8.5 ng/mL	1 ng/mL	GFAAS	Pallidium modifier.	Xiao-Quan et al., 1984

a strongly basic anion exchange column. After elution with sulfur dioxide and evaporation, thallium content was determined by graphite furnace atomic absorption spectrophotometry and differential pulse anodic stripping voltammetry. Results for both techniques were in good agreement. Seawater values were around 15 ng/L.

Oreshkin (1993) reported a direct atomic absorption (AA) method for Tl in sea, river, and aeolian suspended matter collected on membrane ultrafilters. A new two-chamber electrothermal atomizer graphite crucible;ndρaphite cell has been developed. Due to independent multistage heating of a crucible (evaporation zone) and a cell (atomization zone) the processes of matrix decomposition, evaporation, and atomization of vapor of the elements can be controlled and optimized.

To determine thallium in wastewater at the nanogram per milliliter level, Xiao-Quan et al. (1984) used direct injection into a graphite furnace with palladium as a matrix modifier to overcome the interference from halides. Relative standard deviation was 2% and the detection limit was 1 ng/mL.

Luo et al. (1981) used a simple and quick chemical spectrometric method to determine thallium, enriched by means of precipitation from a weakly acidic medium of sodium tetraphenyl boron–soybean protein mixture system. They determined trace amounts of thallium in wastewater. For 1 μg of thallium in 100 mL, standard deviation (STD) was 0.9 with variational coefficient at 9%. For 5 μg thallium in 100 mL, STD was 0.22 and variational coefficient was 4.7%.

7.5.1.2.   Laser-Based Atomic Fluorescence Spectrometry

In general, laser-based techniques possess qualities that allow for highly sensitive and selective analyses. For the detection of Tl, LEAFS in a graphite cup or graphite rod is more than 3 orders of magnitude better than GFAAS (Dougherty et al., 1988). A detection limit of 5×10^{-3} ng/L in aqueous solution was previously found (Falk et al., 1988). By using laser-induced fluorescence in a graphite furnace (LIF-GF), Axner et al. (1993) determined thallium in natural waters. The high sensitivity (ng/L) of this method has eliminated the need for sample pretreatment or preconcentration and therefore has the potential to produce results of good precision and accuracy. There are only small or no matrix interferences.

Baiocchi et al. (1994) used both laser-induced atomic fluorescence spectrometry (LIAFS) and high-resolution inductively coupled plasma mass spectrometry to measure thallium in Antarctic snow. Concentrations were in the order of tenths of picograms per gram. Such extremely low concentrations were ideal for assessing the capabilities of both techniques and evaluating the accuracy of the determination by direct comparison of the results. Both techniques are well suited for direct determination of thallium without preconcentration. In a similar study of snow and firm layers in the Canadian Arctic by LEAFS, Cheam et al. (1996a) reported that surface concentrations of Tl range from 0.3 to 0.9 pg/g, which is a few times higher that those found in Antarctica.

Cheam et al. (1996b) applied LEAFS to study total and dissolved Tl in lake water and pore water. A detection limit of 0.03 ng/L (0.6 fg) was achieved. The method validation was successfully confirmed by using a standard reference material (NIST 1643c) as well as many spike recoveries of digested and undigested unfiltered water samples.

7.5.1.3. Electrochemical Analysis

Jaya et al. (1985) determined thallium in seawater using a potentiometric stripping analysis with chloroacetic acid/sodium chloride as a supporting electrolyte. Concentrations of 10 ng/mL were determined. Miloshova et al. (1994) developed a Tl^+ selective glass sensor for potentiometric measurement of thallium in natural and wastewaters. Thallium concentrations varied from <2 to 75 μg/L. There was no significant interference caused by other metal ions present.

7.5.1.4. Isotope Dilution Mass Spectrometry

Murozumi et al. (1982) used isotope dilution–surface ionization mass spectrometry for the determination of parts per trillion (ppt)(10^{12}) level concentrations of thallium in sea water using a spike solution of ^{203}Tl. The accuracy of the proposed method is excellent in comparison with other methods. The method is applied to the determination of the partitioning of thallium concentrations between particulates and the liquid phase in sea water.

7.5.1.5. Inductively Coupled Plasma Mass Spectrometry

ICP-MS offers sensitivity, accuracy, and speed. It combines many of the best features of the currently approved techniques, including the sensitivity of GFAAS and the rapid multielement sample throughput of inductively coupled plasma atomic emission spectrophotometry (ICP-AES). The quadrupole-based inductively coupled plasma mass spectrometer has been applied to the analysis of many trace elements in natural water, groundwater, and wastewater (Long and Martin, 1991; Pearce, 1991; Beauchemin and Berman, 1989; Barbarino and Taylor, 1987). It can detect in the nanogram per liter range without the need for preconcentration or the introduction of chemical modifiers to eliminate interference effects (pretreatment), and it is thereby a highly accurate technique.

Stetzenbach et al. (1994) used ICP-MS equipped with an ultrasonic nebulizer to determine Tl in groundwater at the nanogram per liter (ppt) level. Detection limit for Tl was 2 ng/L with spring water concentrations of 280–400 ng/L.

7.5.1.6. Inductively Coupled Plasma Atomic Emission Spectrophotometry

For thallium analysis, the sensitivity of ICP-AES is 2 orders of magnitude less than ICP-MS. ICP-AES has a relatively poor detection limit, 60 μg/L for

Tl (van der Jagt and Stuyfzand, 1996), which is far beyond the Tl concentration in waters (2 ng/L)(Stetzenbach et al., 1994) and 6 ± 2 ng/L (Gwendy et al., 1996a). Hence, ICP-AES is seldom used in the study of Tl concentrations in water. Various techniques have been used to enhance the analyte signals in ICP-AES, including ultrasonic nebulization, hydride generation, preconcentration on resin columns, and electrothermal vaporization (ETV). However, these techniques have problems such as matrix effects, overall instrumental complexity, and sample throughput procedures. Pretty and Caruso (1993) reported the use of on-line anodic stripping voltammetry (ASV) for improved detection of lead and thallium in ICP-AES. Thallium was deposited at the electrode at -1.0 V, and a final potential of -0.2 V was used for stripping. Signal enhancements of over 60-fold for 20-mL samples of Tl(I), but no sample results were reported with their studies.

Miyazaki et al. (1992) used ICP-AES combined with solvent extraction to determine thallium in river, mine, and hot spring water. A detection limit of 1.3 ng/mL was obtained. Some samples had Tl concentrations of >15 ng/mL.

7.5.2. Sediment

7.5.2.1. Flame Atomic Absorption Spectrophotometry

Flame atomic absorption spectrophotometry (FAAS) is a useful technique for conducting routine analyses of trace thallium in environmental samples (Table 2). It is a simple, selective, and convenient method. However, because interference with halides and other metals (especially iron in sediments) often occurs, analysis of samples for thallium by flame atomic absorption spectrophotometry usually requires a preconcentration or separation technique or a combination of both.

Ivanova et al. (1994) used a preconcentration/separation technique for the analysis of soils and sediments to improve sensitivity and reduce interferences. Acid digestion of samples using nitric, hydrofluoric, and perchloric acids was carried out in Teflon vessels at 150°C in an autoclave. The digest was cooled, treated with sulfuric acid, and evaporated. Extraction procedure was performed using ammonium tetramethylene carbo-dithiodate/isobutylmethylketone. Iron, which interferes highly with the thallium signal, was reduced with potassium. Thallium concentrations to 0.1 μg/g were determined. Relative standard deviation for soil samples ranged from 4 to 10%.

In a study of geological samples, including stream sediment, Hubert and Chao (1985) used aqua regia/hydrofluoric acid to digest geological reference samples in Teflon beakers. The sample was treated again with a hydrobromic acid/bromine solution. Thallium was extracted with methyliisobutylketone (MIBK). The authors reported no interferences from iron during the procedure because iron is practically unextracted into MIBK at low hydrobromic acid concentrations. Thallium values were in good agreement with reported literature values with detection down to 0.2 μg/g.

Table 2 Summary of Thallium Analysis in Sediments

Sample	Range [Tl]	Detection Limit	Technique	Sample Treatment/Comments	Reference
1. Soil sediments	0.62–1.21 μg/g 2.70 μg/g	0.01 μg/g	FAAS	Acid digestion (H_2SO_4, HF, $HClO_4$) in Teflon vessels (150°C). Digest cooled, treated with H_2SO_4, extracted, and evaporated.	Ivanova et al., 1994
2. Geological samples (includes stream sediments)	0.32–3.9 μg/g	0.2 μg/g	FAAS	Sample digestion in Teflon beakers using aqua regia HF. Treatment with HBr/ Br solution and Tl extraction with MIBK.	Hubert and Chao, 1985
3. River sediments, coal, coal fly ash	0.1–6 μg/g	0.10 μg/g	GFAAS	Sample digestion by pressure deposition using HNO_3, $HClO_4$, and HF. Ascorbic acid to overcome $HClO_4$ interference. Palladium for matrix modification.	Xiao-Quan et al., 1986
4. Sediment soil, coal fly ash	0.55–2.2 μg/g	N/A	GFAAS	Suspension sampling technique in combination with matrix modification (Pd and ascorbic acid).	Zhineng et al., 1987
5. Reference sediments	N/A	N/A	GFAAS	Preconcentration using selective absorption and activated carbon/ potassium xanthogenate. Palladium and ascorbic acid matrix modifiers.	Naganuma and Okutani, 1991

Table 2 (*Continued*)

Sample	Range [Tl]	Detection Limit	Technique	Sample Treatment/Comments	Reference
6. Coal fly ash	0.34–6.2 μg/g	0.1 μg/g	GFAAS	PFTE-lined steel pressure bombs for sample digestion with conc. HNO_3 (coal) and HF (flyash). Part of digest was treated with HBr and ceric sulfate, extracted into di-isopropylether, stripped with ammonium sulfate, acidified with HNO_3, and measured with GFAAS.	Sager, 1993
7. Sediments	0.8 μg/g	1 ng/g	LEAFS	Cold dissolution procedure to dissolve sediment in HNO_3–HF. Also, microwave digestions using aqua regia and HNO_3–HF.	Cheam et al. 1996b
8. Lake Estuary River sediments (Reference Material)	701 μg/kg 963 μg/kg 538 μg/kg	0.1 ng	IDMS thermal ionization	Sample digestion in closed PFTE bombs with HNO_3 and HF and aqua regia (i.e., high pressure digestion).	Waidman et al., 1990
9. BCR CRM 320 (Sediments) German River (Sediments) Elbe Rhine	537 μg/kg 829–1942 μg/kg 1797–5192 μg/kg	N/A	IDMS thermal ionization	As described above.	Waidmann et al., 1992

10. Suspended Matter Marine River Aeolian	1.6–4.2 μg/kg 0.8–0.9 μg/kg 0.83 μg/kg	0.0003 μg	AAS (crucible)	Preliminary pressing of a filter disk carrying sample between layers of graphite powder containing 5–10% S, followed by nonflame atomization of the layered tablet.	Oreshkin et al., 1985
11. Reference sediments	0.65–0.74 μg/g	N/A	ID-ICP-MS	Sample digestion in HF, HNO_3, and $HClO_4$ in Teflon decomposition vessels. Evaporation and further digestions before dilution.	McLaren et al., 1987
12. Sediments Cement kiln dust	0.4–0.9 μg/g	N/A	DPASV	Dried sample digestion in HNO_3 and HF in open Teflon beakers and evaporated to dryness.	Allus and Brereton, 1992
13. Environmental Samples Fly ash Water	0.07–95 μg/g 2.14–7.57 μg/L	0.025 ppm (μg/g)	Differential pulse polarography	Tl is absorbed as its morpholine-4-carbodithioate on microcrystalline naphthalene and then desorbed with 1 M HCl for polarographic analysis. Interferences were eliminated with EDTA [Cd(II) and Pb(II)] and triethanolanine solution [Fe(III)].	Dubey and Puri, 1995

7.5.2.2. Graphite Furnace Atomic Absorption Spectrophotometry

Graphite furnace atomic absorption spectrophotometry (GFAA) was used by Xiao-Quan et al. (1986) for the direct determination of thallium in river sediments, coal, and coal fly ash. They digested samples by pressure decomposition using a mixture of nitric, perchloric, and hydrofluoric acids. Palladium and ascorbic acid were used as matrix modifiers allowing charring at 850°C with no interferences from sample matrices. The purpose of ascorbic acid was to overcome the interference from perchloric acid. According to Xiao-Quan et al. (1986), the suppression effect of perchloric acid on the thallium signal was mainly due to the molecular species of TlCl. Slavin et al. (1981) believed the perchloric acid caused binding of chlorine to analyte atoms, and these were swept out of the absorption tube during atomization and thereby lost from the absorption reading. Xiao-Quan et al. (1986) were able to detect thallium down to 0.10 μg/g using this direct method. Good agreement was obtained for National Bureau of Standard (NBS) reference materials. The relative standard deviation for eight replicate determinations on a river sediment was 2.8%. Zhineng et al. (1987) have developed a method for determining thallium in environmental samples by GFAA using a suspension sampling technique in combination with matrix modification. Palladium and ascorbic acid have been successful in reducing matrix interferences. The authors report good agreement for thallium values in coal fly ash, soil, and sediment samples with certified reference materials.

As with the determination of metals in other samples, Tl determination in sediments by GFAA often improves when a preconcentration procedure is added. Naganuma and Okutani (1991) determined Tl levels in sediments by employing a preconcentration method of selective adsorption. Selective adsorption was achieved using activated carbon and potassium xanthogenate for determining trace amounts of thallium in sediments by GFAA. The sample was combined with activated carbon/potassium xanthogenate, the pH adjusted, mixed with an ultrasonic agitator, then filtered. The activated carbon on the filter paper was dispersed in water with the ultrasonic agitator and injected into the graphite furnace. Both palladium and ascorbic acid were used as matrix modifiers. Results were in good agreement with standard reference materials.

Coal fly ash is a major source of Tl in the environment. Sager (1993) used Teflon-lined steel pressure bombs to digest coal and coal fly ash. Coal samples were digested with concentrated nitric acid. For fly ash samples hydrofluoric acid was added. For thallium determination part of the digest was treated with HBr and ceric sulfate in dilute sulfuric acid to yield trivalent thallium. The thallium was extracted into di-isopropylether, stripped with ammonium sulfite, acidified with nitric acid, and measured by GFAA. Sager (1993) reported a detection limit of 0.1 μg/g.

7.5.2.3. Laser-Excited Atomic Fluorescence Spectrometry

Cheam et al. (1996b) developed a simple procedure to dissolve sediments for LEAFS determination of Tl. They employed a cold dissolution procedure to dissolve sediment in a mixture of HNO$_3$–HF. To further confirm the effectiveness of the proposed dissolution procedure, microwave acid digestions were carried out using two different acid mixtures, aqua regia and HNO$_3$–HF. The method validation was successfully confirmed by using a sediment standard reference material NIST SRM 2704. The detection limit was 1 ng/g of sediment.

7.5.2.4. Isotope Dilution Mass Spectrometry

In an attempt to determine the amount of thallium in a variety of standard reference materials including lake, estuarine, and river sediment, Waidmann et al. (1990) used ID-MS with thermal ionization. Samples were digested in closed PFTE bombs up to 180°C with nitric and hydrofluoric acids. For samples with extremely low concentrations of thallium, low-temperature ashing in an oxygen plasma was carried out. This accommodated samples as large as 1 g. After decomposition electrolysis was carried out, the deposited material was dissolved in nitric/hydrogen peroxide and dried. The dried sample was prepared for loading on a rhenium filament, and thallium was measured using a single-focusing magnetic-type sector field mass spectrophotometer. In most cases standard deviation was <1.5%. Care was taken to heat the rhenium filament slowly as strong heating led to substantial evaporation losses of thallium. Thallium concentrations in standard reference materials were 701, 963, and 538 μg/kg for lake, estuarine, and river sediments, respectively. Based on the mean blank value of 20 blank measurements, the detection limit for this study was 0.1 ng. Using a similar method of ID-MS, Waidmann et al. (1992) determined Tl concentrations in German river sediments. Thallium content ranged from 537 ± 6 to 829 ± 2 μg/kg.

7.5.2.5. Inductively Coupled Plasma Spectrophotometry

Because of its extreme multielement character and low detection limits (0.0005 μg/L) (van der Jagt and Stuyfzand, 1996), ICP-MS is an excellent technique for the determination of trace and ultratrace Tl in sediments. In a comparative study of different digestion procedures using four different analytical methods [ICP-MS, ICP-OES, total reflection X-ray fluorescence spectrometry (TXRF), and instrumental neutron activation analysis (INAA)] for multielement screening of more than 50 elements in sediments of the river Elbe, Krause et al. (1995) reported that thallium could be determined quantitatively only by ICP-MS. The Tl concentration in sediment collected from the Elbe was 1.73 ± 0.07 μg/g. In a study applying a sequential extraction scheme to a suite of marine sediment, marine mud, and soil certified reference materials, Gwendy et al. (1996b) used ICP-MS for the determination of Tl, Cd, Ce, La, Li, Pb, and U. Most values for precision associated with the total element result are in the range of 1–5% RSD. Exceptions arise where concentrations are close to the instrument detection limit, that is, Tl below 0.050 μg/g.

The ICP-MS isotope dilution technique (ID-ICP-MS) can deliver highly accurate data for certain critical applications, such as environmental materials certification. In a study of marine reference materials, Mclaren et al. (1987) applied ID-ICP-MS to analyze trace elements in MESS-1 and BCSS-1 sediments. Samples were prepared in Teflon decomposition vessels and digested in hydrofluoric, nitric, and perchloric acids in a water bath for 3 h. The mixture was evaporated in a Teflon beaker followed by further digestion and evaporation before dilution. Analysis was performed by ID-ICP-MS with RSD at 0.05%. Mclaren et al. (1987) suggested that accuracy and most notably precision are better than those that can be achieved by other ICP-MS calibration strategies, as long as isotopic equilibration is achieved and the isotopes used for the ratio measurement are free of isobaric interferences by molecular species.

7.5.2.6. Electrochemical Techniques

Allus and Brereton (1992) used differential-pulse anodic stripping voltammetry (DPASV) to determine Tl in cement kiln dust and sediment samples. Dry samples were digested with a mixture of nitric and hydrofluoric acids in open Teflon beakers and evaporated to dryness. A $HBr-Br_2$ mixture was added, followed by extraction with diethyl ether to remove Tl^{3+}, which was then reduced to Tl^+ with hydrazine sulfate. Thallium concentrations for the cement dust and sediments ranged from 0.4 to 0.9 $\mu g/g$. Dubey and Puri (1995) used differential pulse polarography to determine thallium in various environmental samples (water, sediments, fly ash). Thallium is adsorbed as its morpholine-4-carbodithioate onto microcrystalline naphthalene and then desorbed with 1 M HCl for polarographic analysis. Detection limit was 0.025 $\mu g/g$ with RSD at 0.95%. Comparison with atomic absorption spectrophotometry (AAS) gave excellent agreement. Interference from Cd(II) and Pb(II) were eliminated with EDTA while that from Fe(III) was eliminated with a triethanolamine solution.

7.5.2.7. Technique Comparison and Instrumental Bias

Kimbrough and Wakakuwa (1994) examined the applicability of different analytical instruments (for instrumental bias) to the analysis of acid digestates of solid materials. They reported the results of data accumulated from 160 environmental laboratories accredited by the state of California to analyze solids for "toxic metals." In their study, a set of four samples were prepared and spiked with 5–4400 $\mu g/g$ in native soil. The ICP-MS data showed very good precision and accuracy, but results were poor at lower concentrations. This was most probably due to spectral problems such as baseline and overlap, which are especially intense below 200 nm (thallium, 190 nm). The main concern is the very low concentration of dissolved solids required for ICP-MS to prevent clogging of the sampling cone. Therefore large dilutions must be made, possibly resulting in dilution and calculation errors seen in the results. Flame AAS results for thallium showed some reproducible positive interferences. Problems with flame AAS are sensitivity and the need for increased sample weight.

7.5.3. Aquatic Plants

According to Smith and Carson (1977), "no data for thallium concentrations in aquatic plants were found." Presently, there are still very few studies on thallium in aquatic plants (Lehn and Schoer, 1985; Flegal et al., 1986; Waidmann et al., 1990).

7.5.3.1. Isotope Dilution Mass Spectrometry

Waidmann et al. (1990) used ID-MS to analyze thallium concentrations in algae reference materials from the German Environmental Specimen Bank. The procedure is briefly described in Section 7.5.4.

7.5.3.2. Spectrofluorimetric Method

Freeze-dried algae and moss were dried to constant weight at 105°C, oxidized by hydrogen peroxide, and then extracted with distilled water (Zitko et al., 1975a). The digest was analyzed for thallium by spectrofluorimetry using a modified method of Kirkbright et al. (1965). The fluorescence was measured on a spectrofluorimeter at an excitation wavelength of 250 nm and an emission of 460 nm.

7.5.4. Fish Tissue

The use of FAAS and graphite furnaces (GFAAS) for the determination of thallium in biological tissues (Table 3) has been recently reviewed by Leloux et al. (1987). Field desorption mass spectrometry, anodic stripping voltammetry, ICP-AES, ICP-MS, and spark source mass spectrometry have all been used to determine thallium levels in biological samples. Of all the techniques, GFAAS is one of the easiest to use on a routine basis and provides a reasonable limit of detection for thallium.

7.5.4.1. Amperometric Titration

In terms of sensitivity, amperometric titration is not comparable to some other analytical methods such as atomic absorption spectrometry, neutron activation, pulse polarography, and anodic stripping voltammetry. However, by employing a microscale device, Hsieh and Ma (1980) found that the sample solution required in amperometry is about 100-fold less than the other methods. Therefore, in terms of the amount of sample used in the analysis, microamperometric titration can be utilized for trace analysis with its accuracy comparable to spectrophotometric methods. Hsieh and Ma (1980) treated the sample by adjusting the concentrated sample solution with pH 9.0 buffer and EDTA masking agent. The masking process must be incorporated into the experimental procedure in order to determine the concentration of the metal

Table 3 Summary of Thallium Analysis in Biological Tissues

Sample	Range [Tl]	Detection Limit	Technique	Sample Treatment/ Comments	Reference
1. Fresh fish	2.6 μg/g	0.2 μg/g (10–5 M)	Amperometric titration	Wet oxidation using H_2SO4 and HNO_3, heating, neutralization with NH_4OH, stepwise extraction with pH 9 buffer, washing, concentration, neutralization of sample solution by NH_4OH, and EDTA as masking agent.	Hsieh et al., 1982
2. BCR 279 Ulva BCR 278 Mussels (I and II, 1981) Homogenized fish	37.9 μg/kg 2.67 μg/kg 15.2–248 μg/kg 4.22 μg/kg	0.1 ng	IDMS thermal ionization	Applied to determine the partitioning of Tl concentrations between particulates and the liquid phase in seawater.	Waidmann et al., 1990

Sample	Concentration	Detection Limit	Method	Procedure	Reference
3. Atlantic Salmon (*Salmo salar*) Toxicity Study	Muscle 114–146 μg/L Liver 80–235 μg/L Gill 27–1430 μg/L	N/A	Spectrofluorometry	Samples digested with H_2SO_4 and H_2O_2, cooled, diluted, sodium hydroxide added, and transferred to separating funnel. Ferric chloride and H_2O_2 were added, Tl extracted with diethyl ether, washed with HCl, and evaporated. Dissolution in HCl. Excit.: 250 nm, emiss.: 430 nm.	Zitko et al., 1975
4. Digestive gland (*Homarus americanus*)	31.22–193.5 ng/g (wet weight)	0.45 ng/g (6 pg/ mL)	ICP-MS	Sample digested in HNO_3 and analyzed by both method of addition and external calibration procedures.	Chou and Uthe, 1995
5. Eggs of winter flounder (*Pleuronectes americanus*)	Below detection limit	<0.05 μg/g	GFAAS	Dried eggs were digested in Teflon lined, high-pressure decomposition bombs using ultrapure concentrated HNO_3.	Vitaliano and Zdanowicz, 1992

ions individually. The concentration limit in the amperometric method is about 0.2 μg/g.

7.5.4.2. Isotope Dilution Mass Spectrometry

Isotope dilution mass spectrometry, which utilizes thermal ionization (a single rhenium filament) as a reliable and highly efficient ionization process, is a very sensitive method of elemental determination. This technique has been applied to determine thallium in reference materials from National Institute of Standards and Technology (NIST), Community Bureau of Reference (BCR) (Table 4), German Environmental Specimen Bank, and other materials (Waidmann et al., 1990). In sample preparation, the sample is decomposed in a Teflon bomb containing a mixture of HNO_3 and HF followed by low-temperature ashing in an oxygen plasma or pressurized decomposition, addition of ^{203}Tl spike solution, and electrolytical deposition prior to ID-MS measurement. The isotopic ratio of $^{205}Tl/^{203}Tl$ is measured at 700°C. The detection limit is 0.1 ng thallium.

7.5.4.3. Spectrofluorimetry

Zitko et al. (1975) analyzed fish samples using a spectrofluorimetric method (Kirkbright et al., 1965). The fluorescence was measured on a spectrofluorometer at excitation wavelength 250 nm and emission maximum 430 nm. Recovery of thallium from the spiked sample was 90%.

7.5.4.4. Inductively Coupled Plasma Mass Spectrometry

Inductively coupled plasma mass spectrometry, which became commercially available in 1983 (Houk, 1985), has been increasingly used for measurement of elements in environmental samples: vegetation, soil, dust, and air (Boomer and Powell, 1987). One advantage of using ICP-MS is its specificity and lower detection limits compared to optical methods.

Table 4   Thallium (Tl) Concentrations in Reference Materials from NSIT, BCR, and the German Environmental Specimen Bank as Determined by Isotope Dilution Mass Spectrometry (IDMS)

Material	Tl Conc. (μg/kg)	Material	Tl Conc. (μg/kg)
Pine needle (NSIT)	0.50[a]	BCR 398 river sediment	538 ± 7
NSIT SRM 1566	4.52 ± 0.21	Algae I 1981	14.3 ± 0.25
Oyster tissue (NSIT)	<5[a]	Algae II 1981	24.8 ± 0.5
BCR 278 Mussel tissue	2.67 ± 0.07	Mussels I 1981	15.1 ± 1.3
BCR 280 Lake sediment	701 ± 9	Mussels II 1981 (Dreissena)	248 ± 8
BCR 277 Estuarine sediment	963 ± 10	Fish, homogenized	4.22 ± 0.10

[a] Value not certified.

Chou and Uthe (1995) developed a method for the determination of Tl and U in the lobster digestive gland samples by ICP-MS. The method is straightforward, with no need for further chemical treatment prior to analysis by ICP-MS. The digest is prepared with nitric acid and analyzed by both method of addition and external calibration procedures. The excellent agreement between both methods suggests that diluted HNO_3 digests of lobster digestive gland contained no material with significant matrix effects on Tl and U measurement.

7.5.4.5. High-Temperature Graphite Furnace Atomic Absorption Spectrophotometry

The use of graphite furnace atomic absorption spectrophotometry for the determination of trace thallium in biological tissues is a most popular and common technique. Vitaliano and Zdanowicz (1992) analyzed thallium concentrations in eggs of the winter flounder (*Pleuronectes americanus*) from Boston Harbor, a contaminated North American estuary. They used graphite furnace atomization with the incorporation of Zeeman effect background correction. They reported that thallium was below the detection limit in all egg samples. To enhance the higher thermal pretreatment temperature during GFAAS atomization for thallium, the addition of palladium and mixed modifiers to fish tissues is recommended (Matsumoto, 1993).

7.5.4.6. Proton-Induced X-ray Emission Spectrometry

Heit et al. (1989) used proton-induced X-ray emission (PIXE) to analyze the thallium concentration in axial muscle of fish tissues (white suckers and yellow perch). The quality assurance of accuracy and precision was determined from the analysis of the NBS 1566 oyster tissue. However, no data were reported for the NBS oyster or the muscle.

7.6. SPECIATION

Interest in the speciation (Table 5) of trace metals has increased over the last decade especially with the knowledge that the chemical form and/or oxidation state of trace metals greatly affects their toxicity and biological function. Speciation of a metal in solution is an important determination because the redox state can dramatically affect the toxicity, adsorption, and transport properties.

Thallium can be found in two different oxidation states, Tl(I) and Tl(III). Although Tl(I) is more stable in aqueous solutions, Tl(III) forms complexes of greater stability. Thallium(I), the lower valency state, is the more toxic. The trivalent state is generally less reactive. In view of its high toxicity even at trace levels, the development of highly sensitive and selective methods to monitor thallium in various samples is of considerable importance. Numerous techniques have been used for thallium analysis. After appropriate sample

Table 5 Summary of Thallium Speciation

Sample	Technique	Sample Treatment	Remarks	Reference
1. Seawater/river water	FAAS	Tl^{3+} complexation with 8-Q column; bromine oxidation to convert Tl^+ to Tl^{3+}.	No effect from seawater; Fe, Cu, and Al were tolerable; 3 ng/mL for 3 min preconcentration time.	Mohammad et al., 1994
2. Soils and minerals	Flow injection, fluorescence spectrophotometry	Tl^{3+} reduced by thiourea to Tl^+.		Tomas et al., 1996
3. River sediments	GF-AAS	HCl/NH_4Cl, HNO_3, and NH_4Cl/CH_3COONH_4 used for leaching. Direct determination but preconcentration required for noncontaminated samples.	AAS signal affected by Ca, Mg, Fe, P, and reagent.	Sager, 1992b
4. Aqueous solution	Radiochemical method	Tl^+ and Tl^{3+} chelated with diethyldithiocarbomic acid.	^{204}Tl for the separation of Tl species based on extraction constant.	Stary, 1987

#	Sample	Method	Description	DL	Reference
5.	Seawater and urine	DPASV-CME method	Tl^{3+} selectively accumulated onto a 8-Q incorporated CME electrode and then reduced Tl^{3+} to Tl by DPASV.	DL[a] at 0.047 ppb (μg/L).	Cai and Khoo, 1995
6.	Aqueous solution	Flow system/DP voltammetry[b]	Tl^{3+} reacts with formic acid by irradiation to produce Tl^{+}	DL at 10 ppb (μg/L)	Barisci and Wallace, 1992
7.	Tap water/ wastewater	Spectrophotometry	Tl^{3+} extracted with amidine/ toluene solution, then react with BG solution.	DL at 18 ppb (μg/L)	Sharma and Patel, 1992
8.	Aqueous solution	ICP-MS	Label Tl with ^{203}Tl, then perform separation of Tl(I) as TlBr precipitates and Tl(III) remains in solution.		Ketterer and Fiorentino, 1995

[a] DL, detection limit.
[b] DP voltammetry, differential pulse voltammetry.

pretreatment, thallium determination has been performed using spectrophotometry, atomic absorption spectrometry, emission spectrophotometry, fluorimetry, laser-induced fluorescence, potentiometric techniques, and stripping voltammetry.

7.6.1. Preconcentration

Ion exchange has been successfully used in the preconcentration of thallium. This approach is normally required for the separation of the analyte from interference species and/or improvements in detection limits. The measurement of Tl concentration is followed by the use of atomic absorption spectrometry, inductively coupled plasma (ICP), optical emission spectrometry, ICP mass spectrometry, and electrothermal atomic absorption spectrometry. The combination of on-line preconcentration with various analytical instruments has been widely and successfully used for the investigation of heavy metals from fresh water (Fang et al. 1984; Beauchemin and Berman, 1989; Malamas et al. 1984) and seawaters (Sturgeon et al., 1981; Willie et al., 1983; Nakashima et al., 1988).

In the study of the complexation chemistry and speciation of thallium for on-line preconcentration with immobilized quinolin-8-ol, Mohammad et al. (1994) examined the effects of various buffering reagents and pH conditions using flame AAS to optimize complexation of Tl^{3+} with 8-Q, immobilized on controlled-pore glass beads. An oxidation method (by bromine) was used to convert Tl^+ to Tl^{3+}. With a flow rate of 6 mL/min, the detection limit of Tl^{3+} is 3 ng/mL, for a 3-min preconcentration time. The major ions in sea water did not interfere with Tl^{3+} preconcentration and the tolerable limits of Fe^{3+}, Cu^{2+}, and Al^{3+} are high enough to permit analysis of river and sea waters.

7.6.2. Flow Injection Spectrofluorimetric Method for Speciation of Thallium

Flow injection offers the possibility of determining species in a mixture without prior separation. Tomas et al. (1996) reported a flow injection method for the determination of Tl(I) and Tl(III) and fluorimetric detection of thallium in flowing solutions. The very simple, inexpensive, and rapid procedure was applied to the determination of Tl in synthetic mixtures, soils, and minerals. Thallium(I) is determined by measuring its fluorescence in an HCl medium (exciting = 227 nm; μm/emission = 419 nm). The determination of Tl(III) is based on the rapid reduction of this ion by thiourea with the concomitant formation of fluorescent Tl(I). Linear calibration graphs were obtained between 5.0×10^{-8} and 1.0×10^{-5} mol/L of Tl(I) and between 5.0×10^{-7} and 1.0×10^{-8} mol/L of Tl(III). The method was successfully applied to the determination of thallium in soils and zinc ores.

7.6.3. Determination of Inorganic and Organic Thallium by Atomic Absorption Spectrophotometry

Huber et al. (1978) demonstrated that thallium undergoes biomethylation in sediment. The DMT was measured using a colorimetric procedure involving prior masking of Tl(I) with EDTA. Curry et al. (1969) developed an atomic absorption method for quantitative determination of inorganic Tl in which Tl(I) and Tl(III) were chelated by sodium diethyldithiocarbamate (NDDC), and extracted into MIBK. The MIBK solution was aspirated directly into an air-acetylene flame and absorbance measured at 276.8 nm. In a study on the differentiation between Tl and DMT species, Morgan et al. (1980) reported that inorganic Tl can be treated with bromine–water to form $TlBr_3$ and can be extracted into MIBK without chelation. They claimed that their method of simultaneous determination of inorganic and organic Tl in a given sample significantly improves the expediency of analysis and the sensitivity.

7.6.4. Speciation of Thallium in River Sediments

Lehn and Schoer (1985) have found a significant part of thallium to be bound to organic detrital, sulfidic or silicatic material, if the total content is at the background level. At contaminated sites, however, major amounts of thallium were in exchangeable forms. Univalent thallium may be nearly as mobile as potassium. It can substitute K^+ and Rb^+ in silicates (Butler, 1962) but is not coprecipitated with amorphous hydroxides (Efremov and Alekseeva, 1957). In the study of freshwater sediment from thallium nonpolluted sites for sequential leaching procedures, Sager (1992b) reported that the sensitivity of graphite furnace AAS can be significantly lowered by matrix effects from both the reagent solutions and also dissolved Ca, Mg, Fe, and P. When the extracts in HCl and NH_4Cl were used in leaching procedures, the charring temperature had to be lowered to 410°C to cope with the volatility of the TlCl species. For noncontaminated samples, direct determination of thallium in the extracts was not sensitive enough, and preconcentration procedures were necessary. As a result of the leaching sequence, Tl is found largely in the nitric acid leachable and the oxalate-leachable fractions. Soluble thallium was completely adsorbed by the sediments from tap water overnight. In subsequent sequential leaching, it was found highly exchangeable at pH 7 against ammonium chloride or ammonium acetate.

7.6.5. Radiochemical Method

Stary (1987) applied ^{204}Tl for the separation studies of thallium species. Thallium(I) and Tl(III) form chelates with diethyldithiocarbamic acid that are extractable into nonpolar solvents. The procedure can be used to concentrate Tl(I) and Tl(III) from large volume samples. The separation of both chelates was achieved by the great difference between the extraction constants (Stary and Freiser, 1978).

7.6.6.   Differential Pulse Anodic Stripping Voltammetry

The great advantage of DPASV is its selectivity of chemical species and its ability of preconcentration of analyte on the electrode. Anodic stripping voltammetry at the mercury electrode is a well-known and established technique for the determination of low levels of metal ions. However, its application to real samples is complicated by interferences from coexisting metal ions and organic compounds present in the sample matrices. In thallium analysis, interfering metal ions include lead, cadmium, copper, bismuth, indium, titanium, and iron. In some cases, the interferences can be eliminated by addition of complexing agents (e.g., ethylenediaminetetraacetic acid and/or surfactants). Another approach to eliminate interferences is the use of chemically modified electrodes (CME) together with medium exchange. The preconcentration step imparts high sensitivity to this technique, whereas the exchange of medium to a pure electrolyte solution removes interferences from the sample background. Cai and Khoo (1995) developed a DPASV-CME method for determination of trace thallium in urine and seawater samples. Thallium(III) was selectively accumulated from a stirred Britton-Robinson buffer solution (pH 4.56) onto a carbon paste electrode incorporating 8-hydroxyquinoline. The ensuing measurement was carried out by DPASV after reducing the Tl(III) to metallic thallium in ammonia–ammonium chloride buffer (pH 10.0). A detection limit of 0.047 μg/L was found for a 2-min accumulation. Effects of the reduction potential, reduction time, and interference from metal ions and organic substances were examined.

Barisci and Wallace (1992) developed a method that allows speciation and electrochemical detection of thallium in flowing solutions. Thallous ions are determined using oxidative detection at a platinum electrode in a sulfuric–formic acid solution. Upon irradiation of this solution, Tl(III) reacts with formic acid to produce Tl(I) that is subsequently detected by differential pulse voltammetry. Discrimination between Tl(I) and Tl(III) can be easily accomplished by carrying out detection with and without photolysis. The limit of detection calculated for a signal-to-noise ratio of 2 at a flow rate of 0.5 mL/min was found to be 10 μg/L for both Tl(I) and Tl(III).

7.6.7.   Spectrophotometric Determination

Sharma and Patel (1992) developed a spectrophotometric procedure for quantification of inorganic thallium [total Tl, Tl(I), and Tl(III)] in tap water, iron ore wash water, and wastewater. Thallium(III) is quantitatively extracted from an aqueous HCl solution with a mixture of toluene–amidine, and the extract is allowed to react with a brilliant green solution. The effect of 10 amidines on the absorbance of the complex were tested, and the most sensitive compound was found to be *N*-(3-toly),-*N'*-phenylbenzamidine. The absorbance of the extract was measured at 640 nm. The detection limit of the method was found to be 18 μg/L in an aqueous solution.

7.6.8. Inductively Coupled Plasma Mass Spectrometry

To measure electron self-exchange rate constants, Ketterer and Fiorentino (1995) used enriched stable isotope labels (^{203}Tl), chemical separations, and ICP-MS to determine the Tl(III)/Tl(I) reaction. The exchange is monitored by mixing the labeled and unlabeled reactants and performing timewise separations through selective precipitation of Tl(I) as TlBr. Isotope abundances are measured in the TlBr precipitate and Tl(III) solution phases using ICP-MS with minimal sample preparation. An NIST 981 (common lead) spike is added as internal standard to correct for mass discrimination. The exchange rate found by their method is in good agreement with those obtained from analogous and radiotracer experiments. They suggested that the method may potentially be extended to the study of other redox couples, which can be suitably prepared in metal-ion-labeled form and for which a suitable separation of the redox forms can be developed.

REFERENCES

Allus, A, and Brereton, R. G. (1992). Determination of thallium in sediments of the River Elbe using isotope dilution mass spectrometry with thermal ionization. *Analyst,* **117,** 1075–1084.

Axner, O., Chekalin, N., Ljungberg, P., and Malmsten, Y. (1993). Direct determination of thallium in natural waters by laser induced fluorescence in a graphite furnace. *Intern. J. Environ. Anal. Chem.,* **53,** 185–193.

Baiocchi, C., Giacosa, D., Saini, G., Cavalli, P., Omenetto, N., Passarella, R., Polettini, A., and Trincherini, P. R. (1994). Determination of thallium in antarctic snow by means of laser induced atomic fluorescence and high resolution inductively coupled plasma mass spectrometry. *Int. J. Anal. Chem.,* **55,** 211–218.

Barbarino, J. R., and Taylor, H. E. (1987). Stable isotope dilution analysis of hydrologic samples by inductively coupled plasma mass spectrometry. *Anal. Chem.,* **49,** 1568–1575.

Barisci, J. N., and Wallace, G. G. (1992). Photoelectrochemical detection and speciation of thallium (I) and thallium (III). *Electroanalysis,* **4,** 139–142.

Batley, G. E., and Florence, T. H. (1975). Determination of thallium in natural waters by anodic stripping voltammetry. *Electroanal. Chem. Interfac. Electrochem.,* **61,** 205–211.

Beauchemin, D., and Berman, S. S. (1989). Determination of trace metals in reference water standards by Inductively-Coupled Plasma Mass Spectrometry with on-line preconcentration. *Anal. Chem.,* **61,** 1857–1862.

Boomer, D. W., and Powell, M. J. (1987). Determination of uranium in environmental samples using inductively coupled plasma mass spectrometry. *Anal. Chem.,* **59,** 2810–2813.

Bowen, H. J. M. (1966). *Trace Elements in Biochemistry.* Academic, New York, p. 241.

Browning, E. (1969). *Toxicity of Industrial Metal,* 2nd ed., Butterworths, London, p. 383.

Butler, J. R. (1962). Thallium in some igneous rocks. *Geokim,* **6,** 514–523.

Cai, Q., and Khoo, S. B. (1995). Determination of trace thallium after accumulation of thallium(III) at a 8-hydroxyquinoline-modified carbon paste electrode. *Analyst*, **120**, 1047–1053.

Canterford, G. S., and Canterford, D. R. (1980). Toxicity of heavy metals to the marine diatom *Ditylum brightwellii*–(West) Grunow: Correlation between toxicity and metal speciation. *J. Mar. Biol. Assoc. U.K.*, **60**(1), 227–242.

Cheam, V., Lawson, G., Lechner, J., Desrosiers R., and Nriagu J. (1996a). Thallium and cadmium in recent snow and firn layers in the Canadian Arctic by atomic fluorescence and absorption spectrometries. *Fresenius J. Anal. Chem.*, **355**, 332–335.

Cheam, V., Lechner, J., Desrosiers, R., Azcue, J., Rosa F., and Mudroch, A. (1996b). LEAFS determination and concentration of metals in Great Lakes ecosystem. *Fresenius J. Anal. Chem.*, **355**, 336–339.

Chou, C. L., and Uthe, J. F. (1995). Thallium, uranium, and $^{235}U/^{238}U$ ratios in the digestive gland of American lobster (*Homarus americanus*) from an industrialized harbour. *Bull. Environ. Contam. Toxicol.*, **54**, 1–7.

Curry, A. S., Read, J. F., and Knott, A. R. (1969). Determination of thallium in biological material by flame spectrophotometry and atomic absorption. *Analyst*, **94**, 744–753.

Davison, R. L., Natusch, D. F., Wallace, J. R., and Evans, C. A. Jr. (1974). Trace elements in fly ash. Dependence of concentration on particle size. *Environ. Sci. Technol.*, **8**, 1107–1113.

de Ruck, A., Vandecasteele, C., and Dams, R. (1988). Determination of thallium in natural waters by electrothermal atomic absorption spectrometry. *Mikrochim. Acta*, **2**, 187–193.

Dougherty, J. P., Costello, J. A., and Michel, R. G. (1988). Determination of thallium in bovine liver and mouse brains by laser excited atomic fluorescence spectrometry in a graphite tube furnace. *Anal. Chem.*, **60**, 336–340.

Dubey, R. K., and Puri, B. K. (1995). Differential pulse polarographic determination of thallium in various environmental samples after adsorption of its morpholine-4-carbodithioate onto microcrystalline naphthalene. *Ann. Chim.* (Rome), **85**, (1–2), 87–95.

Efremov, G. V., and Alekseeva, I. P. (1957). Coprecipitation of thallium with manganese (IV) hydroxide. *Ser. Khim. Nauk*, **15**, 87–91.

Falk, H., Paetzold, H. J., Schmidt, K. P., and Tilch, J. (1988). Analytical application of laser excited atomic fluorescence using a graphite cup atomizer. *Spectrochim. Acta*, **43B**, 1101–1110.

Fang, Z., Ruzicka, J., and Hansen, E. H. (1984). An efficient flow-injection with on-line ion-exchange preconcentration for the determination of trace amounts of heavy metals by atomic absorption spectrometry. *Anal. Chim. Acta*, **164**, 23.

Fangruo, L., and Xiandeng, H. (1994). Determination of sub-ppb levels of thallium in natural water by STPF AAS after preconcentration using a polyurethane plastic foam column. *Atomic Spec.*, **15**(5), 216–220.

Flegal, A. R., Settle, D. M., and Patterson, C. C. (1986) Thallium in marine plankton, *Mar. Biol.*, **90**, 501–503.

Gwendy E. M, Hall, J. E., and Pelchat, J. C. (1996a). Performance of inductively coupled plasma mass spectrometric methods used in the determination of trace elements in surface waters in hydrogeochemical surveys. *J. Anal. Atom. Spectrom.*, **11**, 779–786.

Gwendy, E. M., Gauthier, G., Pelchat, J. C., Pelchat, P., and Vaive, J. E. (1996b). Application of a sequential extraction scheme to ten geological certified reference materials for the determination of 20 elements. *J. Anal. Atom. Spec.*, **11**, 787–796.

Hamaguchi, H., Ohta, N., Onuma N., and Kawasaki K. (1960). Inorganic constituents in biological material. XIV. Contents of thallium, selenium, and arsenic in fish and shells from the Minamata District, Kyushu. *J. Chem. Soc. Jpn.* (Pure Chem. sect.), **81**, 920–927.

Heit, M. (1979). Variability of the concentrations of 17 trace elements in the muscle and liver of a single striped bass, *Morone saxatilis*. *Bull. Envir. Contam. Toxic.*, **23**, 1–5.

Heit, M., Schofield, C., Driscoll, C. T., and Hogkiss, S. S. (1989). Trace element concentrations in fish from three Adirondack lakes with different pH values. *Water, Air, Soil Pollut,* **44,** 9–30.

Houk, R. S. (1985). Mass spectrometry of inductively coupled plasma. *Anal. Chem.,* **58,** 97A.

Hsieh, S. A. K., and Ma, T. S. (1980). Trace analysis of toxic metals. I. Determination of lead, mercury, cadmium and thallium by amperometric titration. *Mikrochim. Acta,* **2**(3–4), 291–300.

Hsieh, S. A. K., Chong, Y. S., Tan, J. F., and Ma, T. S. (1982). Determination of lead, mercury, cadmium, and thallium in foods by amperometry and by atomic absorption spectrometry. *Mikrochim. Acta.* [Wien], **2,** 337–346.

Huber, F., Schmidt, U., and Kirchmann, L. H. (1978). In *Organometals and Organometalloids– occurrence and Fate in the Environment,* Brinkman, F. E. and Bellama, J. M. (eds.), American Chemical Society, Washington, DC, p. 73.

Hubert, A. E., and Chao, T. T. (1985). Determination of gold, indium, tellurium and thallium in the same sample digest of geological materials by atomic-absorption spectroscopy and two-step solvent extraction. *Talanta,* **32,**(7), 568–570.

Ivanova, E., Stoimenova, M., and Gentcheva, G. (1994). Flame AAS determination of As, Cd, and Tl in soils and sediments after their simultaneous carbodithioate extraction. *Fresenius' J. Anal. Chem.,* **348,** 317–319.

Jaya, S., Rao, T. P., and Rao, G. P. (1985). Simultaneous determination of lead and thallium by potentiometric stripping. *Talanta,* **32,** 11, 1061–1063.

Ketterer, M. E., and Fiorentino, M. A. (1995). Measurement of Tl(III/I) electron self-exchange rates using enriched stable isotope labels and inductively coupled plasma mass spectrometry. *Anal. Chem.,* **67,** 4004–4009.

Kimbrough, D. E., and Wakakuwa J. (1994). Interlaboratory comparison of instruments used for the determination of elements in acid digestates of solids. *Analyst,* **119,** 383–388.

Kirkbright, G. F., West, T. S., and Woodward, C. (1965). Spectrofluorimetric determination of microgram amounts of thallium. *Talanta,* **12,** 517–524.

Kogan, B. I. (1970). Rare elements. *Redk. Elem.,* **5,** 27–48.

Korenman, I. M. (1960). *Analytical Chemistry of Thallium.* Israel Program for Scientific Translations, Jerusalem, 1963, Translation of *Analiticheskaya Khimiya Talliya,* Izdatel'stvo Akademii Nauk USSR, Moscow, 1960.

Krause, P., Erbsloh, B., Niedergesab, R., Pepelnik, R., and Prange, A. (1995). Comparative study of different digestion procedures using supplementary analytical methods for multielement-screening of more than 50 elements in sediments of the river Elbe. *Fresenius J. Anal. Chem.,* **353,** 3–11.

LeBlanc, G. A., and Dean, J. W. (1984). Antimony and thallium toxicity to embryos and larvae of fathead minnows (*Pimephales promelas*). *Bull. Environ. Contam. Toxicol.,* **32,** 565–569.

Lehn, H., and Schoer, J. (1985). Thallium transfer from soils to plants: Relations between chemical forms and plant-uptake. 5th Int. Conf., Athens. T. D. Lekkas. (ed.), CEP Consult., Edinburgh, UK. *Heav. Met. Environ.,* **2,** 286–288.

Leloux, M. S., Phu Lich, N., and Claude, R. J. (1987). Determination of thallium in various biological matrices by graphite furnace atomic absorption spectrophotometry using platform technology. *At. Spectrosc.,* **8,** 75–77.

Loehle, C., and Paller, M. (1990). Heavy metals in fish from streams near F-area and H-area seepage basins (U). WSRC-RP-90-482. Westinghouse Savannah River Company, Savannah River Site, Aiken, SC 29808.

Long, S. E., and Martin, T. D. (1991). Determination of trace elements in waters and wastes by inductively coupled plasma mass spectrometry. In *Methods for the Determination of Metals in Environmental Samples.* EPA/600/4-91/010, June.

Luo, X. Y., Zhan, G. Y., and Chang, X. J. (1981). Determination of trace amount thallium in waste water with chemical spectrometry. *Lanzhou Daxue Xuebao,* **3,** 107–113.

Malamas, F., Bengtsson, M., and Johansson, G. (1984). On-line trace metal enrichment and matrix isolation in atomic absorption spectrometry by a column containing immobilized 8-quinolinol in a flow-injection system. *Anal. Chim. Acta,* **160,** 1.

Mathis, B. J., and Kevern N. R. (1975). Distribution of mercury, cadmium, lead and thallium in a eutrophic lake. *Hydrobiology,* **46,** 207–222.

Matsumoto, K. (1993). Palladium as a matrix modifier in graphite-furnace atomic absorption spectrometry of group IIIB–VIB elements. *Anal. Sci.,* **9,** 447–453.

McLaren, J. W., Beauchemin, D., and Berman, S. S. (1987). Application of isotope dilution inductively coupled plasma mass spectrometry to the analysis of marine sediments. *Anal. Chem.,* **59,** 610–613.

Merian, E. (1991). In *Metals and Their Compounds in the Environment.* VCH, Weinheim, p. 1227.

Miloshova, M., Seleznev, B., and Bychkov, E. A. (1994). Chalcogenide glass chemical sensors for determination of thallium in natural and waste water. *Sensors Actuators B,* **19**(1–3), 373–375.

Miyazaki, A., and Tao, H. (1991). Trace determination of thallium in water by laser enhanced ionization spectrometry using electrothermal vaporizer as a sample introduction system. *Anal. Sci.* [Suppl., Proc. Int. Congr. Anal. Sci.], **7**(2) 1053–1056.

Miyazaki, A., Sanpei, T., Tao, H., and Nagase, T. (1992). Determination of trace thallium in natural waters by inductively coupled plasma emission spectrometry combined with solvent extraction. *Nippon Kagaku Kaishi,* **3,** 307–311.

Mohammad, B., Ure, A. M., and Littlejohn, D. (1994). Study of the complexation chemistry and speciation of thallium for on-line preconcentration with immobilised quinolin-8-ol. *Mikrochim. Acta,* **113,** 325–337.

Morgan, J. M., McHenry, J. R., and Masten, L. W. (1980). Simultaneous determination of inorganic and organic thallium by atomic absorption analysis. *Bull. Environ. Contam. Toxicol,* **24,** 333–337.

Murozumi, M., Igarashi, T., and Nakamura, S. (1982). Simultaneous determination of thallium, copper, cadmium and lead in seawater by isotope dilution-surface ionization mass spectrometry. *Nippon Kagaku Kaishi,* **1,** 54–60.

Naganuma, A., and Okutani, T. (1991). Determination of trace amounts of thallium in some sediments by graphite furnace AAS after preconcentration as zanthogenate complex on activated carbon. *Bunseki Kagaku—Japan. Analysts,* **40**(5), 251–256.

Nakashima, S., Ssturgeon, R. E., Willie, S. N., and Berman, S. S. (1988). Determination of trace metals in seawater by graphite furnace atomic absorption spectrometry with preconcentration on silica-immobilized 8-hydroxyquinoline in a flow-system. *Fresenius Z. Anal. Chem.,* **330**(7), 592.

Nriagu, J. O. (1984). *Changing Metal Cycles and Human Health.* Springer, Berlin Heidelberg, New York.

Oreshkin, V. N. (1993). Lowering of detection limits of direct atomic absorption determination of Ag, Bi, In, Tl in sea and river suspended matter. *Okeanologiya* (Moscow), **32**(5), 954–958.

Oreshkin, V. N., Belyayev, Yu, I., and Vnukovskaya, G. L. (1985). Increasing the sensitivity and precision of direct nonflame atomic absorption determination of cadmium, lead, silver, and thallium in marine, river, and aeolian suspended matter. *Oceanology* **25,** No. 6, 794–797.

Pearce, F. M. (1991). The use of ICP-MS for the analysis of natural waters and an evaluation of sampling techniques. *Environ. Geochem. Hlth,* **13,** (222), 50–55.

Pic, P., Ellory, J. C., and Lucu C. (1979). Evidence that K-dependent transport components in addition to Na-K ATPase are involved in Na and Cl excretion in marine teleost gills. *J. Exp. Biol.,* **79,** 1–6.

Pretty, J. R., and Caruso, J. A. (1993). Signal enhancement of lead and thallium in inductively coupled plasma atomic emission spectrometry using on-line anodic strippoing voltammetry. *J. Anal. At. Spectrom.,* **8,** 545–550.

Quinby-Hunt, M. S., and Turekian, K. K. (1983). Distribution of elements in seawater. EOS Trans., *Am. Geophys. Un.,* **64,** 130–131.

Riley, J. P., and Siddiqui, S. A. (1986). The determination of thallium in sediments and natural waters. *Anal. Chim. Acta,* **118,** 117–123.

Sager, M. (1992a). Thallium. In *Hazardous Metals in the Environment,* Stoppler, M. (ed.), Elsevier, Amsterdam, pp. 351–372.

Sager, M. (1992b). Speciation of thallium in river sediment by consecutive leaching techniques. *Mikrochim. Acta,* **106,** 241–251.

Sager, M. (1993). Determination of arsenic, cadmium, mercury, stibium, thallium and zinc in coal and coal fly-ash. *Fuel,* **721**(9), 1327–1330.

Sager, M. (1994). Thallium. *Toxicol. Environ. Chem.,* **45.**

Sharma, M., and Patel, K. S. (1992). Spectrophotometric determination of inorganic thallium in water. *Indian J. Environ. Hlth.,* **34**(3), 219–225.

Slavin, W., Manning, D. C., and Carnrick, G. R. (1981). The stabilized temperature platform furnace. *At. Spectrosc.,* **2**(5), 137–145.

Smith, I. C., and Carson, B. L. (1977). *Trace Metals in the Environment, Volume 1, Thallium,* Ann Arbor Science, Ann Arbor, MI.

Stary J. (1987). Rapid separation of mercury, arsenic, thallium and chromium species. *J. Radioanal. Nucl. Chem.,* **112**(1), 119–123.

Stary, J., and Freiser, H. (1978). *Equilibrium Constants of Liquid-Liquid Distribution Reactions, Part IV, Chelating Extractants.* Pergmon, Oxford.

Stetzenbach K. J., Amano, M., Kreamer, D. K., and Hodge, V. F. (1994). Testing the limits of ICP-MS: Determination of trace elements in ground water at the part-per-trillion level. *Ground Water,* **32**(6), 976–985.

Sturgeon, R. E., Berman, S. S., Willie, S. N., and Desaulnier J. A. H. (1981). Preconcentration of trace elements from seawater with silica-immobilized 8-hydroxyquinoline. *Anal. Chem.,* **53,** 2337.

Tomas, P. R., Martinez-Lozano, C., Tomas, V., and Casajus R. (1996). Simple flow injection spectrofluorimetric method for speciation of thallium. *Analyst,* **121,** 813–816.

Ure, A. M., and Berrow M. L. (1982). In *Environmental Chemistry,* Bowen, H. J. (ed.). Vol. 2. Royal Society of Chemistry, London, pp. 155–195.

van der Jagt, H., and Stuyfzand, P. J. (1996). Methods for trace element analysis in surface water, atomic spectrometry in particular. *Fresenius J. Anal. Chem.,* **354,** 32–40.

Vitaliano, J. J., and Zdanowicz, V. S. (1992). Trace metals in eggs of winter flounder from Boston Harbor, a contaminated North American estuary. *Mar. Pol. Bull.,* **24**(7), 364–367.

Waidmann, E., Hilpert, K., and Stoeppler, M. (1990). Thallium determination in reference materials by isotope dilution mass spectrometry (IDMS) using thermal ionization. *Fresenius' J. Anal. Chem.,* **338,** 572–574.

Waidmann, E., Stoeppler, M., and Heininger, P. (1992). Determination of thallium in sediments of the river elbe using isotope dilution mass spectrometry with thermal ionization. *Analyst,* **117,** 295–298.

Wallwork-Barber, M. K., Lyall, R., and Ferenbaugh, W. (1985). Thallium movement in a simple aquatic ecosystem. *J. Environ. Sci. Hlth.,* **A20**(6), 689–700.

Welz, B., Schlemmer, G., and Mudakavi, J. R. (1988). Palladium nitrate–magnesium nitrate modifier for graphite furnace atomic absorption spectrometry. Part 2. Determination of arsenic, cadmium, copper, manganese, lead, antimony, selenium and thallium in water. *J. Anal. At. Spectrom.,* **3,** 695–701.

Willie, S. N., Sturgeon, R. E., and Berman, S. S. (1983). Comparison of 8-quinolinol-bonded polymer support for the preconcentration of trace metals from sea water. *Anal. Chim. Acta.,* **149,** 59–66.

Xiao-Quan, S., Zhi-neng, M., and Li, Z. (1984). Application of matrix-modification in determination of thallium in waste water by graphite-furnace atomic-absorption spectrometry. *Talanta*, **31**(2), 150–152.

Xiao-Quan, S., Yuan, Z., and Ni, Z. (1986). Determination of thallium in river sediment, coal, coal fly ash and botanical samples by graphite furnace atomic absorption spectrometry. *Can. J. Spectrosc.*, **31**(2), 35–39.

Zhineng, Y., Shan, X.-Q., and Ni, Z. (1987). Determination of thallium and cadmium in environmental samples by graphite furnace atomic absorption spectrometry using suspension sampling technique in combination with matrix modification. *Environ. Chem.*, **6**(3), 32–36.

Zitko, V. (1975a). Toxicity and pollution potential of thallium. *Sci. Total Environ.* **4**, 185–192.

Zitko, V. (1975b). Chemistry, applications, toxicity, and pollution potential of thallium. Fish. Mar. Serv. Res. Dev. Tech. Rep. 518.

Zitko, V., and Carson, W. V. (1975). Accumulation of thallium in clams and mussels. *Bull. Envir. Contam. Toxic.*, **14**, 530–533.

Zitko, V., Carson, W. V., and Carson, W. G. (1975). Thallium: Occurence in the environment and toxicity to fish. *Bull. Envir. Contam. Toxicol.*, **13**, 23–30.

8

ANALYSIS OF THALLIUM IN BIOLOGICAL SAMPLES

Camilo Ríos and Sonia Galván-Arzate

Departamento de Neuroquímica, Instituto Nacional de Neurología y Neurocirugía, Secretaria de Salud, Insurgentes Sur 3877, México D.F., 14269 México

Thallium in the Environment, Edited by Jerome O. Nriagu.
ISBN 0-471-17755-5. © 1998 John Wiley & Sons, Inc.

8.1. INTRODUCTION

Rodenticides and insecticides in which the active agent is a thallium salt have been employed since 1920. Soon thereafter, several cases of acute and chronic thallium poisoning were reported as a result of human ingestion of these products. Thallotoxicosis is still common, not only due to accidental or environmental exposure, but also as a means of self-poisoning and murder (Mulkey and Oehme, 1993).

Symptoms and signs of thallium poisoning are frequently neurological and include neuropathy, chorea, motor disturbances, and coma (Reed et al., 1963; Bank et al., 1972). In some cases of human thallotoxicosis, the nonspecific nature of the clinical markers of thallotoxicosis made it difficult to diagnose (Meggs et al., 1994). Therefore, the differential diagnosis of both acute and chronic thallium poisoning must be assessed by measuring thallium in blood and urine concentrations and, according to Kamerbeek et al. (1971), antidotal treatment with Prussian blue should be applied until thallium urinary excretion of patients does not exceed 0.5 mg of thallium/24 h. Therefore, monitoring of thallium concentrations in body fluids is important, not only for the management of the poisoned patient but also for environmental thallium exposure assessment and for the monitoring of thallium in experimental studies with animals (Ríos and Monroy-Noyola, 1992).

Thallium analysis has been performed by differential pulse anodic stripping voltametry (Liem et al., 1984). Field desorption mass spectrometry (Achenbach et al., 1980), inductively coupled plasma mass spectrometry (Mauras et al., 1993), and furnace atomic absorption spectrophotometry (Chandler and Scott, 1984; Ríos et al., 1989). Some of the early methods for thallium analysis require a previous extraction procedure to avoid interferences or to preconcentrate the element. Direct analysis of thallium has been performed by graphite furnace atomic absorption spectrophotometry and inductively coupled palsma mass spectrometry with sufficient sensitivity. In the case of graphite furnace atomic absorption spectrophotometry, determination is expected to be troublesome since the volatility of the metal restricts the use of high charring temperature for thermal pretreatment with the graphite furnace. About 500°C is the maximum of charring temperature for aqueous thallium standards thermal pretreatment. Important matrix interferences are then expected to appear due to incomplete elimination of organic and inorganic background material, including the always present chloride interference in biological samples. Sulfuric acid has been reported to modify matrix interferences in thallium signal (Slavin et al., 1984), but the sample thermal pretreatment with this matrix modifier is still inadequate (Yang and Smeyers-Verbeke, 1991) for thallium analysis.

The use of a palladium salt, combined with ammonium nitrate (Yang and Smeyers-Verbeke, 1991) is a successful matrix modifier that allows an interference-free thallium determination in urine and blood. However, these reagents are expensive, particularly for experimental or environmental studies in which a large number of samples needs to be processed (Ríos et al., 1989).

Diammonium hydrogenphosphate is commonly used as a matrix modifier for volatile metals such as lead and cadmium (Miller et al., 1987). This reagent is particularly interesting as a matrix modifier because of its low cost. Its action on thallium furnace atomic absorption spectrophotometry signal has not been yet studied in detail. We have extensively used diammonium hydrogenphosphate as a matrix modifier both in experimental toxicological studies as in the clinical diagnosis of thallium intoxication cases in human patients. In this chapter, evidence is shown about the utility of diammonium hydrogenphosphate as a matrix modifier for the direct determination of thallium in blood and urine. This particular technique is also applied to both the analysis of biological fluids and tissue thallium contents.

8.2. EXPERIMENTAL PROCEDURE

8.2.1. Apparatus

Perkin-Elmer model 360 and 3110 atomic absorption spectrophotometers equipped with Perkin-Elmer HGA 2200 and HGA-600 graphite furnaces, deuterium arc background corrector, and Perkin-Elmer AS-60 autosampler were used. A Perkin-Elmer thallium electrodeless discharge lamp operated at 7 W was used at a wavelength of 276.8 nm (slit 0.7 nm). Thallium hollow cathode lamp could also be used. Continuos nitrogen flow served as the purge gas (flow 10 mL/min) for all analysis.

8.2.2. Reagents

Deionized water was obtained from a Milli R/Q water purifier. Suprapur sulfuric acid, Suprapur nitric acid, analytical-grade diammonium hydrogenphosphate, hydrochloric acid (E. Merck, México) and Triton X-100 (Carbiochem) were used.

Stock standard solutions containing 1000 mg/L of thallium was made by weighing 1.303 g of $TlNO_3$, previously dried in an oven at 100°C for 2 h, dissolved in 1 L of deionized water. A working standard solution containing 0.5 mg/L was obtained by proper dilution of the stock solution. Stock solution is stable for at least 2 yr.

Glassware and pipette plastic tips were immersed for several hours in 3% (v/v) concentrated HNO_3/H_2O, thoroughly rinsed in deionized water, and nitrogen gas dried before use. This step is important to avoid cross contamination of the samples with high thallium content (such as tissue samples).

8.2.3. Sample Preparation

Blood samples were obtained by venous puncture using disposable syringes with stainless steel needless and then immediately transferred to heparinized (100 I.U.) polypropylene tubes with caps. Urine samples from patients were acidified to pH = 3.0; all samples were stored frozen until analyzed.

Tissue samples placed in polypropylene tubes were digested in 1–3 mL of concentrated Suprapur nitric acid in a shaking water bath at 60°C for 30 min. This treatment ensures complete destruction of organic matter without thallium loss (Christian, 1969). After digestion, a 100 μL aliquot was taken from the clear solution and diluted (1:5 or 1:10 v/v) with the matrix modifier aqueous solution described below (Ríos et al., 1989).

8.2.4. Matrix Modification

Standard solutions with concentrations ranging from 20 to 150 μg/L were obtained by proper dilutions with different composition solutions. Aliquots of 20 μL were injected into the furnace with a fixed volume Eppendorf pipette in the case of the manual equipment (P-E 360 spectrophotometer). The autosampler for the P-E 3110 equipment also delivered 20-μL samples and standard volumes.

Four matrix modifier solutions were employed to test their effects on thallium standard signals: (a) 1% (w/v) sulfuric acid solution; (b) 1% (w/v) diammonium hydrogenphosphate solution; (c) 1% sulfuric acid plus 1% diammonium hydrogenphosphate combined solution, all of them containing 0.1% (w/v) Triton X-100; and (d) 0.3% g/L of palladium nitrate + 50 g/L of ammonium nitrate dissolved in 0.03 M nitric acid.

8.3. RESULTS

A char temperature study was performed to compare the thallium signal in the presence of the four matrix modifiers. In all cases, atomization temperature was held constant throughout at 2000°C. Results of the first three matrix modifiers are shown in Figure 1. Thallium signal is lost up to 500°C in absence of a matrix modifier (curve A). Sulfuric acid stabilizes thallium signal to 600°C (curve B), and still higher temperature is achieved without loss of thallium in the presence of diammonium hydrogenphosphate (curves C and D). Under the later conditions, thallium begins to volatilize at about 1000°C. It is interesting to note that the thallium signal increases with increasing char temperature up to 500°C in the presence of the phosphate modifier, and this increase is higher for the phosphate–sulfuric acid solution.

Figure 2 shows the thallium signal in the presence of diammonium hydrogenphosphate modifier in comparison to the signal observed with the palla-

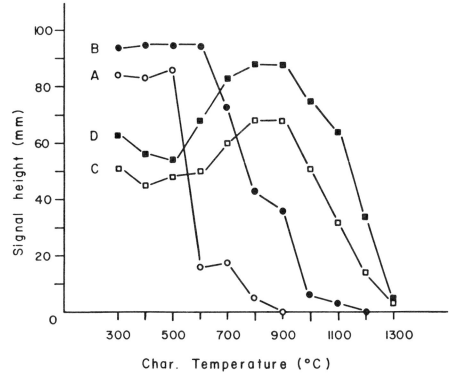

Figure 1. Char study for thallium 50 µg/L standard. Different matrix modifier solutions were used to analyze thallium aqueous standard, as follows. (A) 0.1% Triton X-100, (B) 1% $H_2 SO_4$ and 0.1% Triton X-100, (C) 1% $(NH_4)_2 HPO_4$ and 0.1% Triton X-100, and (D) 1% $(NH_4)_2 HPO_4$, 1% H_2SO_4 and 0.1% Triton X-100.

dium modifier. As clearly observed, palladium modifier stabilizes thallium signal up to 1100°C.

Normal blood samples diluted (a) 5-fold and (b) 10-fold with the diammonium hydrogenphosphate modifier were used to obtain background signal at the thallium resonance line. Note from Figure 3 that background signal decreases with increasing char temperature and reaches a plateau after 900°C. Background signal at 600°C is almost twice the signal at 900°C for 5-fold dilution of blood indicating an incomplete matrix elimination at the lower char temperature. Background signal at 900°C is less than 0.1 absorbance unit, within the compensating range of the deuterium arc background corrector. Therefore, from these particular results it is concluded that 900°C of thermal pretreatment is sufficient to eliminate most of the possible matrix interferences coming from organic matter and chloride.

According to the char studies, temperature settings and other instrumental parameters were chosen as shown in Table 1. As palladium matrix added no

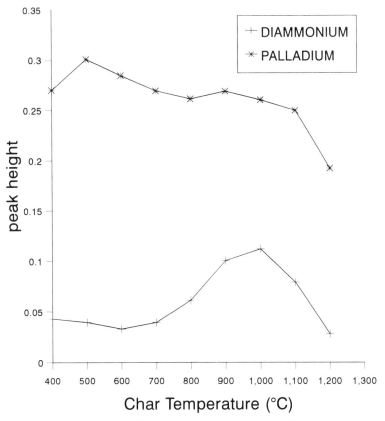

Char Temperature (°C)

Figure 2. Char study for thallium 50 µg/L standard. Thermal stability of thallium signal in phosphate-based matrix modifier is compared with palladium salt modifier signal. +, 1% $(NH_4)_2$ HPO_4 and 0.1% Triton X-100; *, 0.3 g/L of Pd NO_3 + 50 g/L of NH_4 NO_3.

advantage to the thermal pretreatment of thallium samples, we used the diammonium hydrogenphosphate modifiers for the following studies.

Figure 4 shows calibration curves for the matrix modifier solutions. The best sensitivity is achieved in the (A) sulfuric acid medium, as calibration curve slopes shown. Sensitivity in the presence of diammonium hydrogenphosphate is increased by addition of sulfuric acid (C).

8.3.1. Analytical Parameters

Some analytical parameters were obtained in order to compare the two phosphate-based matrix modifiers. Table 2 lists the reciprocal sensitivity for 1% absorption, the detection limits calculated as 2-sigma, and the linear ranges for the calibration curves. The two matrix modifiers have sensitivity and detec-

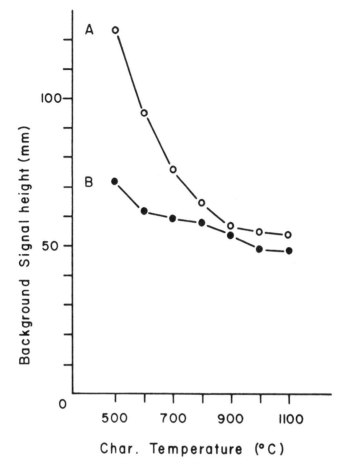

Figure 3. Background signal at 276.8-nm line, as a function of char temperature. (A) Fivefold dilution of blood with 1% $(NH_4)_2$ HPO_4 and 0.1% Triton X-100 and (B) Tenfold dilution of blood with 1% $(NH_4)_2$ HPO_4 and 0.1% Triton X-100.

tion limit values sufficiently low to permit the analysis of thallium concentrations in diluted samples of blood and urine taken from intoxicated patients. The use of phosphate–sulfuric acid solutions is recommended since it permits better thermal stability of thallium with high sensitivity.

The analytical quality control assessment of the method proposed was performed by analyzing a reference material (LabLEADER, Eastman-Kodak Co.) at different dilutions, as shown in Table 2.

The accuracy and precision of the proposed method was also assessed by analyzing 10 times normal blood samples added with 25 and 50 μg/L of thallium resulting in 25.4 ± 3.7 and 51.2 ± 3.9 μg/L of thallium, respectively.

Table 1   Instrumental Parameters

	Phosphate	Phosphate Sulfuric	Sulfuric acid
Dry Temp.°C	100	100	100
Dry Time sec	30	30	30
Char Temp.°C	900	900	600
Char Time sec	30	30	30
Atom. Temp.°C	2000	2000	2000
Atom. Time sec	5	5	5
Inj. volume μl	20	20	20

8.3.2.   Thallium Analysis in Blood and Urine

A simple fivefold dilution of blood samples with the diammonium hydrogen-phosphate solution is proposed for the blood thallium analysis, and a 1:20 dilution with the same matrix modifier solution is proposed for urine thallium determination. When the sensitivity needs to be improved, the addition of sulfuric acid to the ammonium phosphate modifier is recommended.

Table 3 shows typical thallium determination in blood and urine of patients suffering from thallium poisoning, analyzed according to the proposed method (diammonium hydrogenphosphate modifier).

8.3.3.   Thallium Analysis in Tissue

Results of the analysis of thallium in body organ samples of rats treated with a nonlethal (16 mg/kg) thallium administration are shown in Figure 5. A group of animals was perfused through the right ventricle with saline solution to remove loosely bound thallium or thallium coming from the blood.

8.4.   DISCUSSION

The action of sulfuric acid on thallium signal has been ascribed to promoting the elimination of interfering anions, mainly chloride (Slavin et al., 1984). Diammonium hydrogenphosphate, on the other hand, may act by inducing the formation of temperature-stable thallium compounds, perhaps pyrophos-phates, as it has been proposed for other elements (Ríos et al, 1987).

The use of sulfuric acid and diammonium hydrogenphosphate as matrix modifiers for thallium direct determination as proposed here is well-suited for routine analysis of thallium in body fluids of intoxicated patients and body organs of animals in experimental studies (Ríos and Monroy-Noyola, 1992).

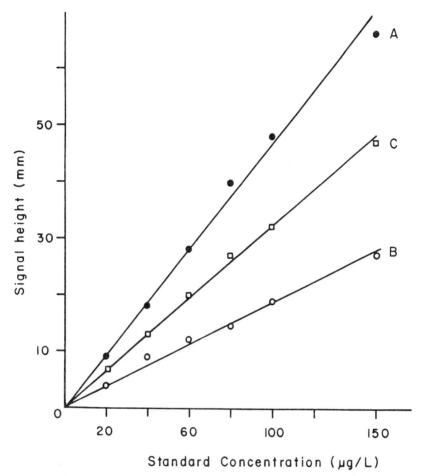

Figure 4. Calibration curves for thallium, prepared in different thallium modifiers. (A) H_2 SO_4–Triton X-100, (B) $(NH_4)_2$ HPO_4–Triton X-100, and (C) $(NH_4)_2$ HPO_4–H_2 SO_4–Triton X-100.

Table 2   Analytical Parameters

	Phosphate	Phosphate Sulfuric	Sulfuric acid
Sensitivity,μg/L	5	8	14.0
Det. Limit, μg/L	4	2	1.5
Linear Range, μg/L	150	150	150.0

Table 3   Typical Thallium Concentrations in Blood and Urine of Thallium-Intoxicated Patients

Patient	Sex	Blood-Tl	Urine-Tl
1	M	0.05 mg/L	0.173 mg/24h
2	M	0.88 mg/L	*
3	F	0.52 mg/L	*
4	F	0.72 mg/L	4.800 mg/24h

* Urine samples were not obtained in patients 2 and 3

As reagents for the analysis of thallium presented here are cheaper and more easily available than the corresponding reagents used in the palladium salts modifier, we strongly recommend their use in large-scale environmental studies. However, an extraction step is needed in the case of measuring "normal" thallium contents in open populations nonexposed to thallium, because of the

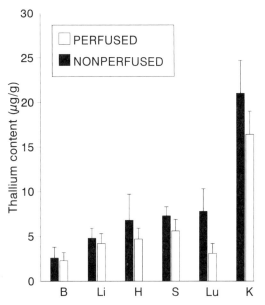

Figure 5. Thallium content in body organ samples of rats treated acutely with 16 mg/kg thallium and sacrificed 24 h after the injection. □, perfused rats, ■, Non perfused rats. B, brain; Li, liver; H, heart; S, spleen; Lu, Lung; K, kidney.

very low content of the metals in such subjects (Paschal and Bailey, 1986). For tissue thallium measurements it is interesting to remark that perfusion of animals with saline solution did not modify significantly the tissue thallium content, indicating that most thallium measured by this way is in the form of tissue-bound metal.

ACKNOWLEDGMENTS

To Jaime Saavedra and Abel Santamaría for useful discussion and critical review. This work was partially supported by CONACYT grants 0935P-M9506 and F256.

REFERENCES

Achenbach, C., Hauswirth, O., Heindrichs, C., Ziskoven, R., Köhler, F., Bahr, U., Heindrich, A., and Schulten, H. R. (1980). Quantitative measurement of time-dependent thallium distribution in organs of mice by field desorption mass spectrometry. *J. Toxicol. Environ. Hlth.*, **6**, 519–528.

Bank, W. J., Pleasure, D. E., Suzuki, K., Nigro, M., and Katz, R. (1972). Thallium poisoning. *Arch. Neurol.*, **26**, 456–464.

Chandler, H. A., and Scott, M. (1984). Determination of low levels of thallium in urine using chelation with sodium diethyldithiocarbamate, extraction into toluene, and atomic absorption spectrophotometry with electrothermal atomization. *Atom. Spectrosc.*, **5**(6), 230–233.

Christian, G. D. (1969). Medicine, trace metals and atomic absorption spectroscopy. *Anal. Chem.*, **41**, 24A–40A.

Kamerbeek, H. H., Rauws, A. G., Ten Ham, M., and Van Heijst, A. N. P. (1971). Prussian blue in therapy of thallotoxicosis. An experimental and clinical investigation. *Acta Med. Scand.*, **189**, 321–324.

Liem, I., Kaiser, G., Sager, M., and Tölg, G. (1984). The determination of thallium in rocks and biological materials at ng g^{-1} levels by differential-pulse anodic stripping voltametry and electrothermal atomic absorption spectrometry. *Anal. Chim. Acta*, **158**, 179–197.

Mauras, Y., Premel-Cabic, A., Berre, S., and Allain, P. (1993). Simultaneous determination of lead, bismuth and thallium in plasma and urine by inductively coupled plasma mass spectrometry. *Clin. Chim. Acta*, **218**, 201–205.

Meggs, W. J., Hoffman, R. S., Shih, R. D., Weisman, R. S., and Goldfrank, L. R. (1994). Thallium poisoning from maliciously contaminated food. *J. Toxicol. Clin. Toxicol.* **32**, 723–730.

Miller, D. T., Paschal, D. C., Gunter, E. W., Stroud, P. E., and D'Angelo, J. (1987). Determination of lead in blood using electrothermal atomisation atomic absorption spectrometry, a L'vov platform and matrix modifier. *Analyst*, **112**, 1701–1704.

Mulkey, J. P., and Oehme, F. W. (1993). A review of thallium toxicity. *Vet. Human Toxicol.*, **35**(5), 445–453.

Paschal, D. C., and Bailey, G. G. (1986). Determination of thallium in urine with Zeeman effect graphite furnace atomic absorption. *J. Anal. Toxicol.*, **10**, 252–254.

Reed, D., Crawley, J., Faro, S. N., Pieper, S. J., and Kurland, L. T. (1963). Thallotoxicosis: Acute manifestations and sequelae. *JAMA*, **183**, 96–102.

Ríos, C., and Monroy-Noyola, A. (1992). D-Penicillamine and prussian blue as antidotes against thallium intoxication in rats. *Toxicology*, **74**, 69–76.

Ríos, C., Valero, H., and Sánchez, T. (1987). Lithium determination in plasma and erythrocytes using furnace atomic absorption spectrophotometry. *Atom. Spectrosc.*, **8**(1), 67–68.

Ríos, C., Galván-Arzate, S., and Tapia, R. (1989). Brain regional thallium distribution in rats acutely intoxicated with Tl$_2$ SO$_4$. *Arch. Toxicol.*, **63**, 34–37.

Slavin, W., Carnrick, G. R., and Manning, D. C. (1984). Chloride investigation in graphite furnace atomic absorption spectrometry. *Anal. Chem.*, **56**(2), 163–168.

Yang, Q., and Smeyers-Verbeke, J. (1991). Effectiveness of palladium matrix modification for the determination of thallium by graphite furnace atomic absorption spectrometry. *Clin. Chim. Acta*, **204**, 23–26.

9

HUMAN
THALLIUM TOXICITY

Guillermo Repetto, Ana del Peso, and
Manuel Repetto

National Institute of Toxicology, P. O. Box 863, 41080 Sevilla,
Spain

Thallium in the Environment, Edited by Jerome O. Nriagu.
ISBN 0-471-17755-5. © 1998 John Wiley & Sons, Inc.

9.1. INTRODUCTION

Unlike other heavy metals that have been featured prominently in toxicological folklore since antiquity, such as lead and arsenic, thallium is a relative newcomer, discovered by W. Crooks in 1861. Since then, thallium has acquired a well-deserved reputation for its toxic properties and is recognized as a potent accidental, occupational, and environmental poison, with incidence in cases of homicide and suicide. However, many aspects of thallium toxicity in human beings, including the toxicokinetics, the mechanisms of action, the teratogenic potential, and the best treatment remain to be elucidated.

Thallium was widely used in the past as a depilatory agent for the treatment of ringworm of the scalp as well as in the treatment of syphilis, gonorrhea, gout, and night sweats in tuberculosis. Although thallium compounds have been banned as pesticides in different countries, thallium is still used as a rodenticide—despite a World Health Organization (WHO) recommendation against such use in 1973 (WHO, 1973; Altuna et al., 1990). In addition, pesticides containing thallium can still be found stored in many homes, causing sporadic poisonings due to their careless handling (Desenclos et al., 1992). Accidental thallium poisoning has resulted from contaminated drugs (Questel et al., 1996) and homeopathic and herbal medicines (Stevens, 1978; Schaumburg and Berger, 1992). Lately, thallium has been employed in myocardial imaging (^{201}Tl) and as a tool in several aspects of biochemistry (Douglas et al., 1990).

Thallium is currently used in industry in the manufacture of lenses and optical components of spectrophotometers made of crystals of Tl-bromoiodide (Chapter 1). It is also used in the manufacture of photoelectric cells with thallium oxysulfide, scintillation counters, fireworks, imitation jewelry, magnesium seawater batteries, semiconductors (Fowler et al., 1993), as well as in an alloy of thallium and mercury in low-temperature thermometers. Other sources of potential poisoning are the dissolutions of thallium malonate and formate employed in mineralogical analysis.

Release of thallium into the environment may occur as the result of human activities, with the emission of vapor and dust, liquid, and solid waste. The main sources of thallium are coal-burning power plants, mining and smelting (Pb, Zn, Cd, Fe, etc.), sulfuric acid production, cement industries, auto emissions, and the agricultural use of phosphate fertilizer. Released environmental thallium affects different trophic levels; human thallium poisoning cases due to the consumption of thallium-contaminated vegetables and fruit grown in the vicinity of cement plants have been reported (Brockhaus et al., 1981).

Thallium produces one of the most complex and serious toxicities known to humans, involving a wide range of organs and tissues. The clinical picture that thallium poisoning presents depends on the time and level of exposure, the rate of absorption, and particularly on individual susceptibility (Prick, 1979; Vergauwe et al., 1990; Mulkey and Oehme, 1993). While as yet unproven, circumstantial evidence suggests quite strongly that thallium produces a basic disorder in energy metabolism, by means of several mechanisms that could be responsible for the pattern of damage seen in nervous tissue, testes and, more rarely, in the heart (Cavanagh, 1991).

Because of the delayed onset of effects, which may vary considerably, and laboratory measurement of thallium in blood and urine being available only through highly specialized reference laboratories, diagnosis of acute or chronic thallotoxicosis is often difficult, except where the etiology is well known; frequently, the source of poisoning is not identified (Prick, 1979; Chandler et al., 1990; Vergauwe et al., 1990; Alarcón-Segovia et al., 1989; Mulkey and Oheme, 1993; Herrero et al., 1995). The cardinal features of gastroenteritis, peripheral neuropathy of unknown origin, and alopecia should call attention to the possibility of thallium poisoning. Unfortunately, the diagnosis of thallium poisoning often occurs only after hair loss is observed (3–4 weeks postabsorption), thus diminishing the effectiveness of treatment and increasing the likelihood of permanent residual effects (Moeschlin, 1980; Rangel-Guerra et al., 1990).

9.2. TOXICOKINETICS

The literature contains many conflicting accounts of the toxicokinetics of thallium, particularly with regard to tissue distribution and rate of elimination. These inconsistencies may result from differences in dose, species, and individual variation (Chandler and Scott, 1986). In the body, thallium moves along potassium pathways involving enteroenteral cycling; without efforts to remove it from the gastrointestinal tract with Prussian blue, it would be recycled repeatedly, thus generating a continuous intoxication.

9.2.1. Cellular Kinetics

The persistent presence of thallium has been explained by similarities in the properties and biological handling of thallium and potassium ions. Since Tl^+ (1.50 Å) and K^+ (1.38 Å) are univalent ions with similar ionic radii, thallium interferes with K-dependent processes and mimics K^+ in its movement and intracellular accumulation in mammals (Gehring and Hammond, 1967). However, once inside the cell, thallium appears to be released less rapidly than potassium. In chronic exposure, the accumulation in tissues accounts for its cumulative toxicity.

9.2.2. Absorption

Whatever the route of administration, the absorption of thallium is rapid and complete, with similar body burden (Lie et al., 1960). The water-soluble salts of thallium (sulfate, acetate, and carbonate) are more rapidly absorbed from the mucous membranes than the less water-soluble forms (sulfide and iodide) (Lund, 1956a,b; Moeschlin, 1980). Thallium can be detected in urine and feces within 1 h, showing plasma level peaks 2 h after oral administration. However, absorption can be delayed due to the constipating effect of the metal (Rauws, 1974).

Chronic accumulation may occur from industrial exposure and as a result of the rapid absorption of thallium through the skin. Percutaneous absorption of thallium can also occur through rubber gloves (Reed et al., 1963). Thallium poisoning has been reported following nasal insufflation of a substance that was believed to be cocaine (Insley et al., 1986). Inhalation of thallium dust (largely in the form of sulfate) occurs either from pyrite burners or from lead and zinc smelters and refiners, as a by-product of cadmium production.

9.2.3.   Distribution

According to Venugopal and Luckey (1978), thallium does not combine particularly well with the serum albumin or other proteins. However, Tl(III) has been found to be a good probe for the glycoprotein, human serum transferrin due to Fe(III) substitution (Bertini et al., 1983).

Some of the toxicokinetic parameters are dose dependent. The volume of distribution in humans is of 3.6–5.6 l/kg, being slightly decreased at high doses (Talas et al., 1983; Talas and Wellhöner, 1983). Evidence for a dose-dependent tissue distribution is supported by experiments conducted with rats by Barclay et al., (1953), who found that, with respect to the other organs, kidney thallium concentration was fivefold higher after administration of tracer amounts than after administration of toxic doses (de Groot et al., 1985).

The distribution of thallium over time shows three phases in an open three-compartment model (Rauws, 1974; Van Kesteren et al., 1980). In the first phase, lasting about 4 h, thallium is distributed through the entire central compartment, composed of the blood and well-perfused peripheral organs and tissues (Rauws, 1974). Thallium is concentrated in blood cells, the blood cell to plasma ratio being 9 (Aoyama et al., 1986). The second phase, from 4–48 h, involves slow distribution throughout the brain, the target organ, and evidences lower concentrations than those in blood. After 24 h distribution is complete, with consequent elimination of thallium from the body. An intensive enteroenteral cycle exists between absorption and secretion, so that the intestine is considered a third compartment (Rauws, 1974). A significant fraction of free plasma thallium crosses the placenta barrier (Stevens and Barbier, 1976).

In experiments using rats, rabbits, and dogs the concentration of thallium in the kidneys was up to 10 times higher than that of any other organ (Barclay et al., 1953; Lie et al., 1960; Talas and Wellhöner, 1983; de Groot et al., 1985). The distribution did not change over time (Lie et al., 1960). However, human thallium distribution is significantly different. The highest concentration of thallium is observed in the heart. Thallium concentration in the kidneys is similar to that in the spleen, the pancreas, the lung, the thyroid, and the stomach. The brain and the fatty tissue showed the lowest levels.

The discrepancy between human case reports and animal studies with regard to the distribution of thallium in the kidneys in comparison to the other organs can be related to interspecies differences or to differences in dose level. These differences in location of thallium suggest caution in extrapolating results obtained in animals to humans (Stokinger, 1981).

Chronic administration causes a similar pattern or distribution, with greater incidence in the kidneys, mainly in the medulla, but with lower concentrations in other tissues than were shown in the acute studies. Retention of thallium in hair follicles and bones increases with age (Smith and Doherty, 1964; Cavanagh et al., 1974; Chandler and Scott, 1986).

Within areas of the rat brain, the highest thallium concentrations were found in the hypothalamus and the lowest in the cortex, independently of the dose (Rios et al., 1989). The implications of the retrograde axonal transport of thallium require further research (Bergquist et al., 1981; Arvidson, 1989).

9.2.4. Metabolism

Inorganic Tl(I) compounds are more stable than the Tl(III) analogs in aqueous solution at neutral pH. In contrast, covalent organothallium compounds are stable only in the trivalent form. Thallium(III) but not Tl(I) can be methylated by methyl-vitamin B_{12} (Huber and Kirchmann, 1978). Thallium-201 chloride, used for single-photon-emission, computarized tomography imaging, eventually decays to mercury (Omura et al., 1995).

9.2.5. Elimination

Thallium is very slowly excreted and may persist for many weeks or even months in the urine and feces of poisoned patients (Repetto and López-Artíguez, 1977). In rats, the ratio of fecal to urinary elimination is approximately 2:1 (Rauws, 1974; Lameijer and van Zwieten, 1977). However, in a human subject poisoned with a lethal dose, the ratio was only 1:4, which was reversed, and approached that found in rats after intensive treatment with Prussian blue (Hologgitas et al., 1980). Fecal excretion may be decreased significantly by paralysis of the small intestine, a characteristic feature of thallium poisoning (Lund, 1956a,b). Thallium concentrations in saliva and milk are, respectively, 15 and 4 times higher than in blood (Richelmi et al., 1980).

9.2.6. Biological Half-life

Due to the large distribution volume of thallium, the elimination half-life is long. The reported estimated half-life in humans is highly variable, with ranges of 1–3 days after low doses (Talas et al., 1983) and 1–1.7 days during intensive clinical therapy after ingestion of a potentially lethal dose (Hologgitas et al., 1980; de Groot et al., 1985). However, other studies gave an elimination half-life of 8 and 30 days (Piazolo et al., 1971). Further toxicokinetic studies of thallium are needed for a better understanding of the behavior of thallium in the human body (Saddique and Petterson, 1983).

9.3. TOXICITY

All thallium compounds are very toxic to humans, but the severity of poisoning depends on the dose, the age of the subject, and the period of exposure. In addition, great differences in its toxic effects occur in accordance with individual susceptibility (Prick, 1979; Vergauwe et al., 1990; Mulkey and Oehme, 1993).

9.3.1. Chemical Speciation

Thallium exists in two oxidation states, Tl(I) and Tl(III). Generally, inorganic Tl(I) compounds are more stable than Tl(III) analogs, but Tl(III) can form stable complexes with many ligands. The element thallium falls into the aluminum family on the periodic table; but monovalent thallium salts are more closely related, chemically and toxicologically, to divalent lead, although the acute toxicity of thallium compounds is much higher (Moeschlin, 1980). Cationic and neutral Tl(III) compounds are present in seawater and fresh water, and some evidence exists suggesting that organic Tl(III) species may originate in the environment during microbial methylation processes (Manzo and Sabbioni, 1988).

The more water-soluble forms of thallium (sulfate, acetate, and carbonate) have been found to be more toxic than the less water-soluble salts (sulfide and iodide) (Gehring and Hammond, 1967). Trivalent thallium compounds are somewhat less toxic than are Tl(I) compounds for different animals, with the exception of guinea pigs, which are equally susceptible to both. In fact Tl(III) will always be reduced to Tl(I) because Tl(III) is unstable in biological systems. In addition, trivalent, but not monovalent, thallium presents high affinity for natural ligands that contain sulfhydryl (-SH) groups (Douglas et al., 1990). Organic compounds of Tl(III) such as thallium dimethyl bromide have been reported to be one-tenth as toxic as thallium acetate when administered to mice (Buschke and Peiser, 1923; Richeson, 1958).

9.3.2. Animal Species and Routes of Exposure

Acute toxicity by oral, subcutaneous, intraperitoneal, and intravenous routes has been determined for 14 inorganic thallium compounds in a total of five animal species. The mean lethal dose (LD_{50}) of the 14 compounds, whether soluble or insoluble, falls into a markedly narrow range, from 15 to 50 mg/kg of thallium, by all routes of administration to all of the various species, indicating that thallium is an extremely hazardous element, independent of the chemical form (NIOSH, 1992). Guinea pigs and rabbits are the most susceptible laboratory animals, while male rats appear to be more susceptible than female rats (Venugopal and Luckey, 1978). The guinea pig demonstrates a toxicologic oddity in that the oral route turns out to be more toxic for it than the intraperitoneal one.

9.3.3. Age

In rats thallium is distributed to the brain more readily in younger animals (Galván-Arzate and Rios, 1994). This suggests that young children may develop thallium neurotoxicity more readily because of an immature blood–brain barrier. However, it has been reported that children tend to be less sensitive to thallium than adults (Prick 1979; Mulkey and Oehme, 1993).

9.3.4. Minimum Lethal Dose

There are few estimates of the minimum lethal dose (LDL0) in humans; reported figures vary considerably from case to case. The average lethal dose for thallium sulfate ranges from 10 to 15 mg/kg (Moeschlin, 1980). Although Gettler and Weiss (1943) proposed a minimum lethal dose of 14–15 mg/kg of soluble salts for an adult, Moeschlin (1980) reported fatalities with 8 mg/kg. Others have reported a lethal dose with as little as 2 mg/kg (NIOSH, 1992).

There have been cases of survival after the ingestion of 650 mg (Grunfeld and Hinostroza, 1964), 1 g (15 mg/kg) (Richelmi et al., 1980), 1.3 g (Grunfeld and Hinostroza, 1964), and 2 g of thallium sulfate (Pedersen et al., 1978), or 20 g of thallium iodide (12 g of thallium) (Koshy and Lovejoy, 1981). Death has been reported after ingestion of 3.2 g of thallium sulfate (Grunfeld and Hinostroza, 1964), 5–10 g of thallium nitrate (Davis et al., 1981), and 10 g of thallous malonate (Aoyama et al., 1986).

9.3.5. Maximum Tolerated Exposure

Studies of human thallium exposure in a wide variety of geographic locations have indicated less than 2 μg as the average daily intake of thallium from environmental sources (Sabbioni et al., 1984). Very little is known about the threshold levels of thallium that may be harmful to health for different age groups. Schaller et al. (1980) have studied 128 employees from production areas in three cement factories. In a group of thallium-exposed subjects urinary thallium was slightly or moderately above the upper normal limit of 1.1 μg Tl/g creatinine. In no cases, however, were clinical effects attributable to observed thallium toxicity.

After the discovery of the contamination accident produced in 1979 at the Lengerich (Germany) cement plant, three consecutive surveys were conducted on a total of 1265 inhabitants from the polluted area. Thallium analysis in urine and hair indicated excessive thallium absorption in more than 60% of the persons examined, with thallium levels 50–100 times higher than those measured in the reference group. The prevalence of subjective symptoms, including headache and sleep disorders, were observed in the exposed subjects. There was no correlation between chemical data and the prevalence of digestive disorders and hair loss (Brockhaus et al., 1981).

Workers in the USSR who had experienced prolonged exposure to thallium carbonate in concentrations exceeding the USSR maximum permissible limit suffered functional changes in their nervous system that were described as "asthenoneurotic" or asthenovegetative syndromes (Shabalina an Spiridonova, 1979). According to another report, minor toxic effects, as evidenced by alopecia areata, were noted in oil refinery workers with urine thallium levels of 10–23 μg/day (Munch et al., 1933).

In the United States and other Western countries the current threshold limit value–time–weighted average (TLV–TWA) for thallium in workplace air is 0.1 mg/m^3, in the context that significant additional skin exposure may be dangerous (EPA, 1992). The former USSR stipulated 0.01 mg/m^3 as the ceiling limit (Shabalina and Spiridonova, 1979). Experimental substantiation of the maximum permissible concentration of thallium carbonate in environmental air has recently been performed (Kamil'dzhanov, 1993).

9.3.6. Tissue Concentrations

Thallium is of natural occurrence as a trace element in human beings. The mean tissue concentration has been calculated as 1.2 μg/kg, from which it was deduced that the thallium content in a 75-kg person would be on the order of 0.1 mg (Weinig and Zink, 1967). Smith and Carson (1977) reported a normal urine level of 1.3 μg/L. Nonetheless, Minoia et al., (1990) established lower reference values of $0.07–0.7$ μg/L. The same authors reported thallium reference values of $0.15–0.63$ μg/L in blood and of $0.02–0.34$ μg/L in serum.

Singh et al. (1975) measured whole-blood thallium concentration in a population of 320 children in New Jersey. Values ranged between less than 5 and 80 μg/L, with a mean concentration of 3 μg/L, but there was no evidence of thallium toxicity in any of the samples analyzed. Over 79% of the blood specimens contained less than 5 μg/L of thallium.

Weinig and Zink (1967) stated that no conclusion on poisoning in forensic practice can be drawn from thallium levels in tissue or urine that are less than 50–100 times their normal values.

Thallium blood concentration has been determined in different case reports (Grunfeld and Hinostroza, 1964; Smith and Doherty, 1964; Repetto and López-Artíguez, 1977; Hologgitas et al., 1980; Nogué et al., 1982–83; Wainwright et al., 1988; Alarcón-Segovia et al., 1989; Vergauwe et al., 1990; Baselt and Cravey, 1995). Crafoord et al. (1996) have recently compiled all the available data from published and unpublished case reports on time-related, high survival, and lethal blood thallium concentrations from cases of acute, human poisoning. The conclusions of this quite detailed study show that most severe poisonings had blood values with a rapid peak around 3 mg/L and a slow decline to 0.1 mg/L values after 9 weeks. Lethal concentrations peak between 2 and 10 mg/L and seem to decline with the same kinetics as the sublethal concentrations, only to be interrupted by death within the period of 12 days.

9.4. MECHANISMS OF ACTION

The precise biochemical mechanisms underlying the clinical manifestations of thallotoxicosis are yet to be confirmed. Many theories have been proposed to account for the toxicity of thallium at the molecular level (Table 1). Circumstantial evidence suggests quite strongly that there is a basic metabolic disorder, which could be responsible for the pattern of damage seen in nervous tissue, skin, testes, and more rarely, in the heart. Thallium induces tissue depletion of the flavoproteins, which may impair the functioning of the metabolic pathways centered on the electron transport chain. As a result the tissues that utilize the most energy, namely the largest cells of the body, the mitotically active and most actively contractile, simply begin to run out of energy (Cavanagh, 1991). It is likely that several mechanisms are operative in the cell.

9.4.1. Potassium Substitution

Thallium substitutes for potassium in many biochemical reactions due to the similarity of the ionic radii of both metals, which are 0.147 and 0.133 nm, respectively. Thallium produces alterations in normal physiological processes such as neurotransmission and muscle excitability (Mullins and Moore, 1960). Tachycardia, hypertension, and certain digestive disorders have been interpreted as neurogenic dysfunctions related to toxic vagus nerve damage, but direct action of thallium on the myocardium as well as on vascular and smooth intestinal muscle has been documented (Hasan and Ali, 1981).

Several K^+-dependent proteins present higher affinity for Tl^+ than for K^+ (Britten and Blank, 1968; Williams, 1970). Pyruvate kinase (PK) is a glycolytic

Table 1   Main Mechanisms of Thallium Toxicity

K substitution
Alteration of K-dependent processes
Muscle excitability, neurotransmission
K-dependent proteins
Pyruvate kinase (PK) inhibition
Na/K-ATPase inhibition
Enzyme stimulation
Protein synthesis inhibition
SH and other group affinity
Enzyme inhibition: SDH, MAO, catepsin, etc.
Mitochondrial swelling
Keratin alteration
Riboflavin depletion/thiamine deficiency
Oxidative stress
Lipid peroxidation
Glutathione depletion
Enzyme inhibition
Amino acid and neurotransmitter alteration
Calcium interaction

enzyme complex strongly inhibited by Tl^+ at higher concentrations, possibly due to the formation of Tl^+–ADP complex (Kaye, 1971). Thallium binds Na^+–K^+-ATPase with 10-fold higher affinity than K^+ (Gehring and Hammond, 1967; Britten and Blank, 1968), with inhibition at high concentrations (Inturrisi, 1969). This leads to the accumulation of thallium within the cell at the expense of potassium, and impairs osmotic regulation, the generation of the electrochemical potential gradient responsible for the electrical excitability of nerve cells, and the generation of free energy (Mulkey and Oehme, 1993). Na^+-K^+-ATPase inhibition causes mitochondrial swelling and vacuolization, which are common electron microscope findings in thallium-poisoned neurons (Spencer et al., 1973). Active mitochondrial accumulation of thallium also uncouples mitochondrial oxidative phosphorylation (Melnick et al., 1976). Although thallium does not readily accumulate in lysosomes (Ando and Ando, 1990), some type of interaction has been recently proposed (Repetto and Sanz, 1993).

At low levels, thallium activates other K^+-dependent enzymes, such as phosphatase, homoserine dehydrogenase, vitamin-B_{12}-dependent diol dehydrogenase, L-threonine dehydratase, and AMP deaminase (Smith and Carson, 1977). Thallium also causes progressive destabilization and irreversible damage to ribosomes, inhibiting protein synthesis and preventing keratinization, which, in turn, contributes to alopecia (Cavanagh et al., 1974; Hutlin and Naslund, 1974).

9.4.2. Affinity for SH and Other Groups

Trivalent, but not monovalent, thallium, presents high affinity for natural ligands that contain sulfhydryl (-SH) groups (Douglas et al., 1990). These groups are important in several classes of enzymes, such as hydrolases, oxidoreductases, and transferases (i.e., flavoenzymes, pyridoxal phosphate-dependent enzymes, and thiol proteases). Succinic dehydrogenase, monoamine oxidase, and the protease cathepsin are very potently inhibited, in parallel with the specific increase in the striatal protein content (Hasan et al., 1977b). The possible implications of this thallium biochemical effect still remain to be clarified.

The inactivation of sulfhydryl groups is also responsible for increasing the permeability of mitochondria, leading to water influx and swelling (Spencer et al., 1973). Thallium also blocks formation of the numerous disulfide bonds between cysteine residues in keratin, which is manifested clinically by alopecia and anomalies in nail growth (Mee's lines) (Thyresson, 1951). However, thallium inhibition of acid phosphatase and guanine deaminase, both enzymes that do not depend on sulfhydryl groups for activity, suggests that binding can occur at other groups.

Moreover, thallium compromises mitochondrial energy production by inhibiting pyruvate dehydrogenase complex (PDC), succinate dehydrogenase (SDH) (Truhaut, 1958), and even cytosolic lactate dehydrogenase (LDH)

(Repetto et al., 1994). Thus, the metal cation blocks the catabolism of carbohydrates and the entry of electrons into the electron transport chain, thereby decreasing ATP generation via oxidative phosphorylation (Melnick et al., 1976). The presence of ketones in the urine and clinical findings of metabolic acidosis in thallotoxicosis are a consequence of the inhibition of normal carbohydrate metabolism.

9.4.3.   Riboflavin Depletion

Thallium precipitates riboflavin and impairs cell energy metabolism by causing a reduction in riboflavin and a depletion of flavine adenine nucleotides and flavoproteins that are required in pyruvate metabolism and the electron transport chain (Kuhn et al., 1933; Aliev and Gasanov, 1965; Cavanagh and Gregson, 1978; Cavanagh, 1979a,b). Many features of thallium toxicity are similar to those found in riboflavin deficiency, including peripheral neuropathy, loss of hair and dermatitis (Schoental and Cavanagh, 1977). In addition, a common pathogenetic mechanism of damage from thiamine deficiency (beriberi) is reflected by the sensitivity of the testes to thallium intoxication, which produces significant inhibition of sperm and changes in the structure of Sertoli cells (Formigli et al., 1986). The testes are peculiarly dependent on glucose as a substrate for energy metabolism, exactly as are nerve cells and skin appendages. A less striking feature is the involvement of heart muscle, which may also be related to this tissue's high energy utilization. Damage to heart muscle is frequent in chronic thiamine deficiency (Swank et al., 1941), and myonecrosis has been seen in thallium intoxication in humans (Prick et al., 1955; Cavanagh et al., 1974; Prick, 1979; Davis et al., 1981).

9.4.4.   Oxidative Stress

Thallium increases lipid peroxidation in the rat brain (Hasan and Ali, 1981). This situation of oxidative stress is probably due to the reported decrease in several enzymes (Hasan et al., 1977a,b) and to thallium's affinity for sulfydryl groups, which may reduce glutathione levels (Chandler and Scott, 1986). In fact, the antioxidant effect of selenium antagonizes thallium-induced free radical/peroxide generation and counteracts its toxicity in vivo (Canther, 1974).

Ali et al. (1990) found thallium-induced alterations in amino acids (glutamic acid, aspartic acid) and neurotransmitters (glutamic acid, dopamine, serotonin) in the brain of rats exposed acutely or subacutely to the metal ion. The interaction with central neurotransmitter systems may at least in part explain the origin of extrapyramidal manifestations, the increase in the spontaneous discharge rate of Purkinje neurons in rat cerebellum occurring during thallium poisoning (Truhaut, 1958), the high urinary level of catecholamine products in thallium poisoning, and the favorable response to the application of L-dopa in the treatment of choreiform sequelae of human thallium poisoning (Hasan and Ali, 1981).

Thallium may increase intracellular Ca^{2+} by several mechanisms: uncoupling oxidative phosphorylation, or disrupting the normal antioxidant processes in the cell, and including inhibition of Na^+-K^+-ATPase and F_0/F_1-ATP synthase. Thallium depolarizes membranes (Mullins and Moore, 1960) and antagonizes the effect of calcium on the heart (Komulainen and Bondy, 1988).

Different arylthallium(III) compounds are analogues of organomercurials and produce selective inhibition of a number of enzymes (e.g., lactate dehydrogenase, phosphoglycerate kinase) and site-specific modifications of protein and t-RNA (Douglas et al., 1990).

Radiation injury and mercury deposits in internal organs as a result of ^{201}Tl chloride used for single-photon emission, computarized tomography imaging, and the potential injury due to radiation exposure during long-distance flights have recently been reported (Omura et al., 1995). Thallium-201 may produce mild radiation injury, causing a change of L-amino acids to D-amino acids, with a peak 2 weeks after administration and lasting for several weeks. Moreover, the metal eventually decays to mercury, another very toxic element (Omura et al., 1995).

9.5. ACUTE EFFECTS

The ingestion of any thallium salt is followed, after a latency period, by the gradual development of symptoms that are often difficult to relate to thallium toxicity without chemical evidence. Although the presentation of gastroenteritis, polyneuropathy, and hair loss are pathognomonic for thallium poisoning, the signs and symptoms are quite variable, involving many organs and systems.

Even though the severity is dose dependent (Prick et al., 1955; Shabaline and Spiridonova, 1979), the onset of thallium poisoning is usually insidious, reaching a maximum in the second or third week after exposure, followed by a very slow recovery or death (Grunfeld and Hinostroza, 1964; Rauws, 1974; Arena, 1976; Cavanagh, 1979a). Four stages can be characterized (Moeschlin, 1980; Lovejoy, 1982; Rangel-Guerra et al. 1990) (Table 2), as described in the following sections.

9.5.1. Immediate (3–4 h)

The target organ in this period is the gastrointestinal tract. Ingestion of thallium may give rise to vomiting but often only nausea is experienced. Abdominal pain, anorexia, diarrhea, and hematemesis may occur (Chandler and Scott, 1986; Moeschlin, 1980).

9.5.2. Intermediate (Hours to Days)

When large doses are taken, paraesthesia, lethargy, delirium, myocardial abnormalities, convulsion, and coma appear soon after ingestion and are followed by central respiratory failure and death within a few days (Munch, 1934; Prick

Table 2   Essential Symptoms of Thallium Poisoning

Gastrointestinal alterations
Deposits of pigment in hair roots (Widy's Lines) (early sign)
Retrosternal diffuse pain
Abdominal cramps
Refractory constipation
Thirst and central sleeplessness
Hysteriform behavior
Polyneuritis, chiefly of the lower extremities
Tachycardia
Loss of hair from the second week onward
Lunula strips on the fingernails (Mee's stripes)

et al., 1955; Grunfeld and Hinostroza, 1964; Cavanagh et al., 1974; Shabalina and Spiridonova, 1979).

In less severe cases, the onset of symptoms may be insidious. For the first 3 or 4 days the patients usually feel relatively well, except for a refractory constipation. A characteristic early sign is dark pigmentation around the hair root, which is already present after the fourth day (Moeschlin, 1980). After a latency period of about 3–7 days, the main target organ becomes the nervous system in which there is a gradual development of hyperaesthesia (sensation of walking on felt), paraesthesia, hyperalgesia of the lower limbs (affecting especially the soles of the feet), followed by motor weakness of the lower limbs and foot dropsy. Encephalopathy and cranial nerve involvement, with ptosis, ophthalmoplegia, retrobulbar neuritis, or facial paralysis occur in severe poisoning. At the same time, or later, characteristic retrosternal and diffuse abdominal pain appear; other characteristic symptoms are excessive thirst and intractable insomnia. Further mental changes occur, which are often interpreted as hysteria.

9.5.3. Late (2–4 Weeks)

In the second week there is a gradual development of tachycardia, usually associated with a moderate systolic hypertension. The more persistent and the more pronounced the increase in pulse rate, the worse the prognosis. Progression of toxic polyneuritis is shown by the loss of the exaggerated tendon reflexes and the development of complete areflexia of the lower limbs and occasionally of the arms as well. Initial hyperaesthesia is replaced by hypoaesthesia; and peroneal paralysis may develop, in some cases accompanied by atrophy of the musculature. At times the arms and cranial nerve tissues are affected. Development of psychotic behavior, with hallucinations and dementia, has also been reported (Prick et al., 1955).

Along with the development of severe polyneuritis, at the beginning or more rarely at the end of the second week, the main target organ becomes the skin, and the characteristic symptom of hair loss appears. By the third week there is usually complete alopecia. The skin becomes dry and scaly following damage to the sebaceous glands. A period of hyperhydrosis is followed by anhydrosis and sweat gland destruction, which may lead to acne. During the third or fourth week semilunar white stripes may appear on the fingernails (Mees's lines), a sign of impaired growth; the stripes grow toward the free edge, but they may not become apparent until the second month when they have grown clear of the nail bed (Prick, 1979).

9.5.4. Residual (Months)

Thallium poisoning recovery requires several months and may not be complete (Munch et al., 1933; Prick et al., 1955). The most common sequelae are giddiness, lack of concentration, intense headache, memory failure, and emotional disturbances—often associated with a gradual decay of intelligence (Manzo and Sabbioni, 1988). The most common findings in poisoned children are mental abnormalities (retardation and psychosis), abnormal reflexes, ataxia, and tremor (Reed et al., 1963). Hair loss is usually reversible, but severe thallium poisoning may lead to permanent alopecia. Some months after poisoning, pronounced caries may become apparent and severe muscular atrophy develops.

9.6. CHRONIC EFFECTS

Subacute and chronic intoxication are associated with cumulative thallium retention in tissue. As with lead, the subacute syndrome may develop without a recognized acute illness. Any sudden release of thallium from these tissue stores may precipitate acute toxic symptoms as in lead poisoning. Thallium is a cumulative poison, and a fulminating lethal outcome was observed in a patient who took a second overdose after a failed suicide attempt (van Kesteren et al., 1980; Rasmussen, 1981).

A high degree of variability is present in chronic thallotoxicosis symptomatology. Although the effects of chronic thallium poisoning may resemble those of acute poisoning, symptoms may be nonspecific and thallium intoxication may not be suspected until alopecia occurs. Excitation and insomnia are often the initial symptoms. After exposure for weeks or months, characteristic joint pain (such as in the ankles, knees, and thoracic spine), weakness, and sometimes polyneuritis occur (Prick et al., 1955). The hair may fall out after a few months. Anorexia, vomiting, weight loss, depression, hysterical laughter, fever, cardiac disturbances, dryness of skin with lack of perspiration, and albuminuria have also been noted.

Moeschelin (1980) reported three cases of homicide due to chronic poisoning. The most consistent findings were severe polyneuritis, with inability to walk, amaurosis, and pronounced cachexia. In two cases there was only slight hair loss. There was notably poor resistance to infection, suggesting some impairment of immunity.

In another case reported by Schmidbauer and Klingler (1979), the clinical picture was dominated by a polyneuropathy, a lesion of the optic nerve and psychiatric abnormalities. A particular feature was early loss of sensitivity of the anterior rami of the intercostal nerves. The typical symptoms of thallium toxicity, such as pain, skin dryness, constipation, and insomnia, did not occur. Significant abnormalities remained 8 years after the intoxication.

Polyneuritic symptoms, sleep disorders, headache, asthenia, skin abnormalities, hair loss, and gastrointestinal dysfunction were found to be the major health effects associated with the consumption of thallium-contaminated vegetables and fruit grown in the vicinity of a cement plant (Brockhaus et al., 1981).

A survey of 51 Soviet Union workers chronically exposed to thallium revealed a functional nervous syndrome of asthenia and neurosis, or asthenia, autonomic dysfunction, and vascular disorders (Stokinger, 1981).

Antitumoral effects of thallium against N-ethyl-N-nitroso-urea-induced tumors have been reported (Barroso-Moguel et al., 1994a). In addition, thallium has been shown to produce decreases in the activity of some drug-metabolizing enzymes dependent on cytochrome P-450; the metal should, therefore, be capable of interfering with the metabolism of organic carcinogens (Fowler et al., 1993).

9.7. ORGANS AND SYSTEMS AFFECTED

Target organs are the gastrointestinal tract, the peripheral and central nervous system, and the skin. However, the signs and symptoms are quite variable, involving many organs and systems; for example, in severe poisoning, the patient may die early due to myocardial failure.

9.7.1. Gastrointestinal Tract

In almost all cases the patient initially suffers from nausea and vomiting (Prick et al., 1955; Schaumburg and Berger, 1992; Feldman and Levisohn, 1993; Moore et al., 1993; Tabandeh et al., 1994). During the 3–4 days following thallium exposure, the patient generally feels well; at the end of the first week the patient complains of nonspecific abdominal pain. The characteristic sign in this initial stage is obstinate constipation, which fails to respond to the usual purgatives (Moeschlin, 1980). In severe cases there is usually hypoacidity or anacidity of the gastric juices, and severe gastrointestinal bleeding may occur (Prick et al., 1955; Moeschlin, 1980; Saddique and Peterson, 1983; Meggs et al., 1994).

9.7.2. Nervous System

Peripheral neuropathy is typically caused by thallium poisoning in humans. Clinical and pathological alterations indicate that the longest sensory and motor nerve fibers are affected first, while shorter nerve fibers, such as those proximal to the limbs and in the cranial nerves, may be affected several days later (Cavanagh et al., 1974). Neurological symptoms usually appear in 2–5 days in acute exposure cases, which are characterized by a painful, rapidly progressive peripheral neuropathy that dominates clinically in the second or third week (Bank, 1980; Vergauwe et al., 1990; Desenclos et al., 1992; Schaumburg and Berger, 1992; Tabandeh et al., 1994; Niehues et al., 1995; Meggs et al., 1994). Sensory disturbances include pain and paraesthesia of the lower limbs, numbness in the fingers and toes, with loss of pin-prick and touch sensation (Prick, 1979). Occasionally, hyperaesthesia involving the soles of the feet and tibial region occurs; the mere weight of bed sheets on the lower extremities may cause excruciating pain (Cavanagh et al., 1974). Initial hyper-reflexia then vanishes and complete arreflexia of the lower extremities ensues (Prick et al., 1955; Cavanagh et al., 1974; Moeschlin, 1980). Motor neuropathy is manifested by muscle weakness and atrofia, which is always distal in distribution. The lower body extremities are primarily affected and cranial nerve participation is rare (Desenclos et al., 1992; Moore et al., 1993). Unlike Guillain-Barre syndrome, stretch reflexes are preserved until late in the course of toxicity, and prominent sensory changes precede motor weakness. The conduction velocities of faster fibers significantly decrease in an early stage of acute thallium poisoning and recover following recuperation from the poisoning (Dumitru and Kalantri, 1990; Araki et al., 1993).

Central nervous system impairment causes mental changes, with psychosis, disorientation, hallucinations, dementia, and deterioration in intellectual function (Thompson et al., 1988). Hysterical behavior may also be seen by the end of the first week following ingestion (Prick et al., 1955; Cavanagh et al., 1974). Excessive thirst and intractable sleeplessness appear to be due to a central nervous system effect of thallium rather than to the severe pain that patients may suffer. Insomnia, headache, anxiety (Cavanagh et al., 1974), irritability, restlessness (Feldman and Levisohn, 1993; Herrero et al., 1995), tremor, ataxia, and choreoathetosis may be produced (Smith and Doherty, 1964; Moeschlin 1980). Loss of perception of pain, temperature, vibratory and position sense, and an unsteady gait may all develop (Schaumburg and Berger, 1992; Tabandeh et al., 1994).

In very serious acute poisonings and commonly in chronic cases true "pseudobulbar paralysis" due to peripheral neuritis of the cranial nerves is observed, with paralysis of the ocular muscles, ptosis, facial paralysis, amblyopia, and paralysis of the recurrent nerve. Shortly before death, paralysis of the vagal nerve may supervene, this possibly being the direct cause of death. Lethargy, coma (Desenclos et al., 1992), convulsions (Feldman and Levisohn, 1993), and cerebral edema with central respiratory failure can also occur (Moeschlin, 1980).

The autonomic nervous system is also altered in the second week, evidencing a gradual development of tachycardia due to direct vagal nerve damage along with a moderate increase in blood pressure, fever, salivation, and sweating (Paulson et al., 1972).

Many types of sequelae have been reported, including mental retardation, psychosis, abnormal reflexes (Dumitru and Kalantri, 1990), ataxia, and tremor (Reed et al., 1963; Wainwright et al., 1988), lasting even more than 30 years (Barnes et al., 1984).

9.7.3. Skin

Alopecia is the most characteristic sign of thallium poisoning and usually appears at the end of the second week. In the third week, there is nearly complete alopecia. This long latent period coincides with the maturation period of the new epithelial cells of the hair papilla. In most cases, facial, pubic, and axillary hair are spared, but they may also disappear (Buschke and Peiser, 1923; Munch, 1934; Prick et al., 1955; Cavanagh et al., 1974; Herrero et al., 1995; Feldman and Levisohn, 1993; Desenclos et al., 1992; Moore et al., 1993).

Other findings are an early anhydrosis—due to destruction of the sweat glands—which may be preceded by an increase in sudation. The skin takes on a remarkably dry, slightly scaly appearance. These signs are almost certain evidence of thallium poisoning, along with a distinct blackish pigmentation, which appears as early as 3–4 days after thallium exposure, inside the hair root. If poisoning has occurred by repeated doses, several zones of pigmentation may be found (Prick et al. 1955; Moeschlin, 1980). Partly necrotic acne can also be seen. Dystrophy of the nails, with the appearance of white semilunar bands (Mee's lines), appears 3–4 weeks after intoxication. Maculopapular rash that progressed to scabbing has also been reported (Tabandeh et al., 1994). Several months after thallium exposure, pronounced caries of the teeth may become apparent (Moeschlin, 1980).

9.7.4. Cardiovascular System

In the second week there is usually gradual development of tachycardia (Wainwright et al., 1988; Moore et al., 1993; Herrero et al., 1995). This may be due to the stimulating effect of thallium on ATP in chromaffin cells, leading to an increased output of catecholamines and direct myocardial damage by thallium or to a direct effect on pacemakers. From the electrocardiographic changes, a direct toxic lesion of the myocardium is more probable (Moeschlin, 1980; Rangel-Guerra et al., 1990), showing changes similar to those observed during hypokalemia: flattening of the T-waves in the limb leads, and, occasionally, even inversion in leads II and III; accompanied by flattening or inversion of the T-waves in the chest leads, more often in V_2 than in V_4 (Moeschlin, 1980; Niehues et al., 1995). Hypertension (Wainwright et al., 1988; Herrero et al., 1995; Feldman and Levisohn, 1993; Desenclos et al., 1992; Meggs et al.,

1994), cardiac arrhythmias, bradycardia (Roby et al., 1984), and angina have been reported.

9.7.5. Eye

Severe bilateral optic neuritis was reported in 25% of acute thallium poisoning cases and in virtually all cases of chronic poisoning. In patients with optic neuropathy, the electroretinogram may be abnormal as soon as 2 days after the intoxication, and visually evoked responses delayed (Shamshinova et al., 1990; Moore et al., 1993; Tabandeh et al., 1994). The toxic action of thallium seems to occur mainly in the retina, possibly resulting in an ascending (retinal) atrophy of the optic nerve (Hennekes, 1983; Labauge et al., 1991). Impaired color vision, decreased visual acuity, optic atrophy, retrobulbar neuritis with central scotoma, ophthalmoplegia, nystagmus, nearly complete blindness and lens opacity (Grant, 1986; Rangel-Guerra et al., 1990; Moore et al., 1993; Tabandeh et al., 1994) have been described.

9.7.6. Reproductive and Developmental Effects

Experimental evidence suggests that the reproductive system is highly suscepti- ble to thallium. Decreased libido and impotence was noted in humans who suffered chronic exposure to the metal (Mulkey and Oehme, 1993). The testes accumulated high levels of thallium. Moreover, morphological and biochemi- cal changes in the testes and decreased epididymal sperm motility were noted in rats exposed to 10 ppm of thallium in drinking water (Formigli et al., 1986). Thallium has been shown to cross animal and human placenta (Ziskoven et al., 1980). However, teratological research in mammals (rats, mice, and cats) has produced conflicting results (Gibson and Becker, 1970; Bank, 1980; Manzo and Sabbioni, 1988).

A significantly greater number of malformations than expected was noted in a human population exposed long-term to thallium-contaminated vegetables produced in the vicinity of a cement plant (Dolgner et al., 1983). However, no specific pattern of congenital malformation was found in the children examined. Maternal exposure to thallium caused skin and nail dystrophy, alopecia, and low body weights in the newborn children of thallium-intoxicated mothers (Moeschlin, 1980). Moreover, there are a number of case reports of human thallium intoxication during pregnancy in the literature, following which no congenital anomalies were observed (Sikkel et al., 1959; Petersohn, 1960; van Maarseveen, 1962). Stevens and Barbier (1976) reported six cases of thallium poisoning in the first trimester of pregnancy in which no consequent anomalies were detected in the newborn. Even abortion failed to occur when thallium was taken for this purpose.

9.7.7. Other Organs and Systems

In the blood, hemolytic changes, anemia, leucocytosis, eosinophilia, lymphope- nia, and thrombocytopenia have been reported (Symmonds, 1953; Reed et al., 1963; Cavanagh et al., 1974; Saddique and Peterson, 1983; Luckit et al., 1990). In some patients, however, the iron in the sera may be slightly increased. Disturbances in porphyrin metabolism have also been found (Moeschlin, 1980).

Respiratory failure (Desenclos et al., 1992), pleuritic chest pain (Meggs et al., 1994), muscle tenderness, and connective tissue disease (arthralgia, polyarthritis) have been reported (Alarcón-Segovia et al., 1989).

Liver damage occurs but is not prominent clinically. Mild elevations in transaminases (SGOT) were reported in two patients who snorted thallium sulfate (Insley et al., 1986). Renal involvement is rare, despite the fact that the kidneys accumulate the highest thallium concentration of any organ. Albuminuria, with erythrocytes, leucocytes, and casts (Moeschlin, 1980), and elevated blood urea nitrogen (Meggs et al., 1994) can be seen.

Autoimmune alterations of connective tissue have been described (systemic lupus erythematosus with positive antinuclear antibodies) (Alarcón-Segovia et al., 1989; Herrero et al., 1995) and a notably poor resistance to infection (Moeschlin, 1980).

9.8. PATHOLOGICAL FINDINGS

Pathological studies of humans with thallium intoxication have been limited. Thallium causes severe lesions in skin cells, hair follicles, and the glanda of the skin, including the total atrophy of the follicles and their replacement by scar-tissue collagen or fat, and keratinization inhibition. Microscopic examination of scalp hair roots shows a tapered or bayonet anagen with the presence of black pigmentation at the base in 95% of the cases 4 days after ingestion of thallium (Saddique and Peterson, 1983; Feldman and Levisohn, 1993).

Thallium ions may cause direct damage to the intestinal epithelium and a variety of pathological and functional changes have been observed (Manzo and Sabbioni, 1988). Lymph follicle centers show distinct lesions, which indicate the toxic effect of thallium on lymphopoyesis. Histological changes in the capillaries are present; in the central nervous system (CNS), the cortical vessels may be engorged (Stokinger, 1981). Central lobular necrosis of the liver, with liver cell swelling and damage to the renal tubular epithelium, can be produced; but the hepatorenal lesions per se are rarely lethal. Most changes are localized in the renal medulla (Appenroth et al., 1995). Iritis, lid inflammation, cataracts, and retinal damage have also been reported (Smith and Carson, 1977; Manzo and Sabbioni, 1988). Electron microscopy of cardiac and skeletal biopsies revealed lipid droplets, increased sarcoplasm, and widening of some of the tubules (Niehues et al., 1995).

Thallium produces a distal and predominantly sensory neuropathy, with some involvement of the central nervous system (Manzo and Sabbioni, 1988). As indicated by the extensive experience of Cavanagh and colleagues (Cavanagh, 1991), the usual patterns of clinical, electrophysiological, and pathological damage in the peripheral nervous system are those of the dying-back type, with the longest fibers, both sensory first and motor later, suffering more injury than short fibers. Preferential damage to the large fibers of sensory neurons also occurs. The relatively slow evolution of the toxic process is dose dependent, since high doses produce rapid development (Davis et al., 1981; Cavanagh, 1991).

Peripheral nerves display distal axonal degeneration, chromatolysis in many of the nerve cells, and secondary myelin loss (Cavanagh et al., 1974; Kennedy and Cavanagh, 1977). Axonal swelling appears to be an early change, as occurs in in vitro studies (Spencer et al., 1973). Electron microscopy studies of sural nerves from individuals with thallium neuropathy have yielded widely divergent findings. One report has illustrated axonal mitochondria swelling and vacuolization similar to those seen in experimental studies (Spencer et al., 1973). The other described nonspecific myelin changes (Bank et al., 1972).

Another topographical point arose from the occurrence of chromatolysis in anterior horn and dorsal root ganglion cells as well as early alterations in the neurons of cranial nerve motor nuclei. It was noteworthy that chromatolysis was significantly more advanced in lumbosacral than in cervical cord levels (Kennedy and Cavanagh, 1977). Analogous changes in the lateral sympathetic chain substantiate the significance of sympathetic stimulation and damage for a number of the symptoms of thallium poisoning. The Goll bundles are also damaged. There are no lesions of central cerebral nerve nuclei, except for retrograde degeneration of the dorsal nuclei ganglion cells of the vagus (Kazantzis, 1986).

Munch and co-workers (1933) performed autopsies on six patients with fatal thallium poisoning. Localized areas of edema and vascular engorgement were found in the cerebral hemispheres and brainstem, while chromatolytic changes were prominent in motor cortex neurons, third nerve nuclei, substantia nigra, and globus pallidus.

9.9. LABORATORY FINDINGS

9.9.1. Thallium Identification and Quantification

Since thallium is radiopaque, an abdominal X-ray was proposed as useful in detecting its presence in the liver and in documenting adequate gastric decontamination following acute ingestion (Grunfeld and Hinostroza, 1964).

A definitive diagnosis and monitoring treatment require analysis of body fluids or excreta for thallium. Today the most accurate technique used for measurement of thallium is atomic absorption spectrophotometry (Amore, 1974; Wakid and Cortas, 1984). An even more sensitive analytical method is neuron activation, by which the thallium content in a single hair can be measured. Many cases of thallium poisoning are not diagnosed because laboratory measurement of thallium in blood and urine is only available through highly specialized reference laboratories (Alarcón-Segovia et al., 1989; Vergauwe et al., 1990), and the source is not identifiable (Chandler et al., 1990). The specimens of choice are urine and whole blood, but hair, saliva, feces, stomach contents, and gastric lavage fluid can also be used, mainly for forensic purposes.

Thallium is excreted in the urine for many weeks following exposure (Chandler et al., 1990). In severe intoxications, urinary excretions have been greater than 5–10 mg/24 h (Grunfeld and Hinostroza, 1964; van der Merwe, 1972; Pedersen et al., 1978; Koshy and Lovejoy, 1981). The most reliable test is a 24-h, urine-quantitative assay. Normal values are 0–5 μg/L (CDC, 1987). Collection of aliquot samples should be simple in patients who undergo forced diuresis treatment. Minoia et al. (1990) established reference values in urine of 0.07–0.7 μg/L. The value of the thallium mobilization test is debatable, since urinary excretion has been reported to be identical before and after potassium administration (Insley et al., 1986).

The thallium reference values are 0.15–0.63 μg/L in blood and 0.02–0.34 μg/L in serum (Minoia et al., 1990). Blood levels above 300 μg/L indicate severe ingestion. In another study, 8% of blood specimens on routine examination contained less than 5 μg/L of thallium (Singh et al., 1975).

Concentrations in saliva may be 15-fold higher than those in urine (Richelmi et al., 1980), and thus may serve as a measure of thallium toxicity in people with severe constipation, renal failure, or other conditions that prevent collection of standard biological materials. Estimates of thallium concentrations in the normal population are 7–15 ng Tl/g hair.

9.9.2. Microscopic Examination of Hair Roots

Microscopic examination of scalp hair roots shows tapered or bayonet anagens, with the presence of black pigmentation in 95% of the cases 4 days after ingestion of thallium (Saddique and Peterson, 1983; Feldman and Levisohn, 1993).

9.9.3. Biochemical Investigation

A slightly depressed serum potassium level and mild hypochloremic acidosis have been observed, probably reflecting thallium-induced renal tubular dysfunction (Reed et al., 1963; Cavanagh et al., 1974). Characteristic changes of hypokalemia may appear on the electrocardiogram (Paulson et al., 1972). Renal malfunction is demonstrated by diminished creatinine clearance, raised blood urea nitrogen, and proteinuria (Prick et al., 1955; van der Merwe, 1972; Cavanagh et al., 1974; Hologgitas et al., 1980). Increased urinary excretion of catecholamine metabolites has also been described.

Thallium is hepatotoxic and increases alkaline phosphatase, aspartate transaminase, and alanine transaminase activities. Increased bromosulphthalein

(BSP) retention times have been reported in severe poisoning (Papp et al., 1969; Cavanagh et al., 1974; Arena, 1976). During the acute phase, the urinary excretion of porphyrin and porphyrin precursors may be markedly increased, as occurs in lead poisoning.

9.9.4. Hematological Investigations

There are no specific laboratory findings associated with thallium poisoning. Anemia and hemolytic changes are occasionally reported, but the lymphocytosis and eosinophilia sometimes seen are probably due to secondary infection (Bachevitch and Prokoptchouk, 1929; Prick et al., 1955).

9.10. DIAGNOSIS AND DIFFERENTIAL DIAGNOSIS

Although rare, thallium poisoning still occurs. The diagnosis of thallotoxicosis is often difficult unless the etiology is well known. A tentative diagnosis of thallium poisoning can be made on clinical criteria. The cardinal features of gastroenteritis, peripheral neuropathy of unknown origin, and alopecia should alert the clinician to the possibility of thallium poisoning. Early symptoms of extremely painful paraesthesia, with exquisite tenderness to touch appearing within 24–48 h of ingestion, support the diagnosis. Unfortunately, the diagnosis of thallium poisoning often occurs only after hair loss is observed (3–4 weeks postabsorption), thus diminishing the effectiveness of treatment and increasing the likelihood of permanent residual effects. It must be stressed that hair loss does not always occur (Bank et al., 1972; Hologgitas et al., 1980; Rangel-Guerra et al., 1990). Furthermore, early treatment on the basis of clinical signs and symptoms is recommended.

Radiographs may reveal metallic densities in thallium-contaminated food and in the liver area. Microscopic examination of scalp hair roots shows the presence of black pigmentation in 95% of the cases 4 days after ingestion of thallium (Saddique and Peterson, 1983). Laboratory measurement of thallium in serum and urine is only available through highly specialized reference laboratories, and laboratory confirmation of poisoning may be delayed.

Differential diagnosis should by performed with acute intermittent porphyria, Guillain-Barre syndrome, systemic lupus erythematosus, botulism (Bertrán-Capella et al., 1972), pellagric vitamin deficits, diabetic polyneuritis, and polyneuritis caused by toxic agents, including arsenic, lead, gold, ethylalcohol abuses, hydrocarbons, CO, and triorthocresylphosphate.

9.11. TREATMENT

Treatment methods in thallium poisoning are highly controversial. The lack of an effective antidote to thallium poisoning is reflected in the wide variety of treatments that have been employed (Prick et al., 1995; Reed et al., 1963; Schaller et al., 1980; Nogué et al., 1982–83; Lehmann and Favari, 1985; Rios and Monroy-Noyola, 1992; Meggs et al., 1994; Barroso-Moguel et al., 1994b). The treatment objective in thallium poisoning is to enhance the elimination of the metal from the body. This is accomplished by inhibiting absorption/ reabsorption (enterohepatic cycle) of the metal from the digestive tract and by mobilizing the thallium from tissue storage sites without exacerbating the symptomatology. Due to the high amount of fixed thallium in the body, all elimination procedures have to be used for a number of days (5–7).

Decontamination should be started by emesis induction and/or gastric lavage and the administration of Prussian blue (PB). Given orally, this compound absorbs monovalent thallium cations in its crystal lattice, thereby interrupting enterohepatic circulation. Prussian blue and the Tl–PB complex are not absorbed from the gastrointestinal tract and are eliminated in the feces. About 7% of PB is degraded to cyanoferrate, which is absorbed and rapidly eliminated in the urine, carrying some bound thallium with it in the process. Prussian blue has a high therapeutic index and is, for all practical purposes, of very low toxicity (Lehmann and Favari, 1984). It is most effective when administered within the first 48 h after ingestion, but results from clinical and experimental studies suggest PB has therapeutic utility throughout the course of acute and chronic intoxications (Heydlaur, 1969; van der Merwe, 1972; Stevens et al., 1974; Barbier, 1974; van Kesteren et al., 1980; Sabbioni et al., 1982; Lehmann and Favari, 1984; Rios and Monroy-Noyola, 1992; Meggs et al., 1994; Barroso-Moguel et al., 1994b). Therapy is continued at least until less than 0.5 μg/24 h are excreted in the urine, according to the symptomatology (Saddique and Peterson, 1983).

Two chemical species of Prussian blue are used: the so-called insoluble form (ferric ferrocyanide) and the colloidal soluble form [potassium ferric hexacyanoferrate(II)]; the latter form, preferably given intraduodenally, is more effective but is not commercially available (Stevens et al., 1974). It has been reported that the best antidotic capacity is achieved by Prussian blue with low crystalline size (Kravzov et al., 1993). Though Prussian blue is not approved for human use by the U.S. Food and Drug Administration, it is available in the United States as a laboratory reagent.

Multiple-dose activated charcoal may be used particularly in areas where Prussian blue is not readily available, but it is not as effective (Lund, 1956a,b; Kamerbeek et al., 1971; Meggs et al., 1994). The recommended dose of activated charcoal is 500 mg/kg day, given twice daily.

Other chelating agents such as sodium diethyldithiocarbamate (dithiocarb), ethylenediaminetetraacetic acid (EDTA), diethylenetriaminepentaacetic acid, D-penicillamine, dimercaprol (BAL), thiouracil, and diphenyldithiocarbazone (dithizone) are not clearly effective or are even contraindicated due to side effects or the formation of lipophilic chelates with thallium, which are redistributed and cause more neurological damage (Hurlbut et al., 1990; Careaga-Olivares and González-Ramírez, 1995). Although D-penicillamine should not be used alone, the combination of Prussian blue and penicillamine potentiated the protective effect of the former in experimental animals (Rios and Monroy-Noyola, 1992; Barroso-Moguel et al., 1994b).

Administration of high doses of riboflavin to try to counteract metabolic effects of thallium has also been proposed (Cavanagh, 1991). Cathartics such as cisapride may be effective in improving gastric emptying and relieving constipation (Vrij et al., 1995).

Potassium chloride diuresis increases thallium excretion by K blocking tubular reabsorption of thallium, or due to the mobilization of thallium from intracellular stores (Gehring and Hammond, 1967; Papp et al., 1969; Bank et al., 1972; Paulson et al., 1972). As it can cause dangerous redistribution of thallium to the central nervous system, potassium chloride should be used with caution and requires close monitoring of renal function and serum potassium levels. After a few days the final outcome is usually beneficial (Cavanagh et al., 1974). However, the combination of potassium chloride and Prussian blue should not be used (Meggs et al., 1994).

Forced diuresis using furosemide and mannitol enhances the elimination of thallium (Nogué et al., 1982–83) and may be useful in acute intoxication. Hemofiltration was found ineffective in eliminating thallium, and hemodialysis is only indicated in the initial stages of acute intoxication, when thallium plasma levels are relatively high, or in the case of renal failure (Saddique and Peterson, 1983; Wainwright et al., 1988).

Charcoal hemoperfusion has been reported to be the most effective procedure for eliminating absorbed thallium, if used within 48 h of thallium ingestion, during the distribution phase (de Groot et al., 1985). The side effects of thrombocytopenia and hypogammaglobulinemia must, however, limit its use.

9.12. PROGNOSIS

Thallotoxicosis is a serious illness with high morbidity and mortality, the outcome of which is hard to estimate (Mulkey and Oehme, 1993). Prognosis depends on the dose of poison ingested, on the interval between the intake of poison and the onset of treatment, and on individual sensitivity, which varies considerably. In general, cases with a fulminating onset are rapidly fatal due to pulmonary or cardiac failure (Aoyama et al., 1986). The prognosis is unfavorable in the presence of persistent tachycardia, severe polyneuritis, paresis, and particularly where serious central nervous system symptoms such as progressive dementia and pronounced mental changes predominate (Moeschlin, 1980).

In all instances, recovery requires several months and may be incomplete. Chandler et al. (1990) reported a case that required 95 days of assisted ventilation and 224 days of hospitalization. In a follow-up study conducted 4 years after thallium intoxication, Reed et al., (1963) found a 58% incidence of chronic neurologic defects in surviving patients involving both the peripheral and central nervous systems. Long-term sequelae of thallium poisoning include insomnia, tremors, visual disturbances, peroneal paralysis, and psychiatric disorders (Manzo and Sabbioni, 1988). Severe and presumably permanent deterioration in intellectual function (memory and performance abilities) has been documented in at least one case (Thompson et al., 1988). However, if thallium poisoning is recognized and treated early, the chance for full recovery is good (Mulkey and Oehme, 1993).

9.13. PREVENTION

The acceptable daily intake (ADI) has not yet been established. The use of thallium salts as rodenticides should be banned universally. In the prevention of thallium poisoning at the workplace, substitution by less toxic compounds should be advocated wherever possible. Where thallium compounds must be used, strict safety precautions should be observed to prevent ingestion, inhalation, or skin contact. Food, drink, and cigarettes should be prohibited at the worksite. Personal hygiene should be ensured; protective clothing and respiratory protective equipment should be worn where inhalation of thallium-containing dust may occur (Kazantzis, 1986).

The current TLV–TWA for soluble thallium compounds in workplace air is 0.1 mg of Tl/m^3 in the United States and other Western countries (Manzo and Sabbioni, 1988; Sax, 1989; ACGIH, 1995). This was based mainly on analogy with other highly toxic metals, and considered that significant additional skin exposure may be dangerous. A 40-h weekly exposure to an airborne thallium concentration of this magnitude would correspond to a thallium concentration in urine of about 100 $\mu g/L$ (Marcus, 1985). The Soviet Union adopted a maximum allowable concentration (MAK) of 0.01 mg/m^3 of thallium for the soluble salts, bromide, and iodide (1972 list). All workers in industry exposed to thallium should periodically have routine urine screens for thallium. Surprisingly, a fatal case of industrial thallium poisoning has never been reported in the literature (Saddique and Peterson, 1983).

Marcus (1985) proposed an alert level, which would require action, of 50 $\mu g/L$ of thallium. From measurements of airborne thallium together with biological monitoring, Marcus considered the present hygiene standard of 0.1 mg/m^3 to be acceptable.

REFERENCES

ACGIH(SM). (1995). *1995–1996* Threshold Limit Values (TLVs(TM)) for Chemical Substances and Physical Agents and Biological Exposure Indices (BEIs(TM)). Am. Conf. Govt. Ind. Hyg., Cincinnati, OH.

Alarcón-Segovia, D., Amigo, M. C., and Reyes, P. A. (1989). Connective tissue disease features after thallium poisoning. *J. Rheumatol.,* **16,** 171–174.

Ali, S. F., Jairaj, K., Newport, G. D., Lipe, J. W., and Slikker, W. Jr. (1990). Thallium intoxication produces neurochemical alterations in rat brain. *Neurotoxicology* **11,** 381–390.

Aliev, A. M., and Gasanov, A. S. (1965). Reaction of some salts of rare and rare earth elements with group B vitamins. *Chem. Abstr.,* **62,** 5142.

Altuna, E., Marti, J., Azparren, J. E., and Bustillo, J. M. (1990). Thallium poisoning (in Spanish). *Rev. Clin. Esp.,* **186,** 362.

Amore, F. (1974). Determination of cadmium, lead, thallium and nickel in blood by atomic absorption spectrometry. *Anal. Chem.,* **46,** 1597–1599.

Ando, A., and Ando, I. (1990). Accumulation of each element in the lysosome of tumor and liver cell as systematized according to its location in the periodic table. *Appl. Radiation Isotopes,* **41,** 327–331.

Aoyama, H., Yoshida, M., and Yamamura Y. (1986). Acute poisoning by intentional ingestion of thallous malonate. *Human. Toxicol.,* **5,** 389–392.

Appenroth, D., Gambaryan, S., Winnefeld, K., Leiterer, M., Fleck, C., and Braunlich, H. (1995). Functional and morphological aspects of thallium-induced nephrotoxicity in rats. *Toxicology,* **96,** 203–215.

Araki, S., Yokoyama, K., and Murata, K. (1993). Assessment of the effects of occupational and environmental factors on all faster and slower large myelinated nerve fibers: A study of the distribution of nerve conduction velocities. *Environ. Res.,* **62,** 325–332.

Arena, J. M. (1976). *Poisoning.* Charles C Thomas, Springfield, IL, pp. 141–143.

Arvidson, B. (1989). Retrograde axonal transport of metals. *J. Trace Elements Exper. Med.,* **2,** 343–347.

Bachevitch, M., and Prokoptchouk, A. (1929). Paychose epileptiforme apres acetate de thallium. *Ann. Dermatol. Syph.,* **10,** 383–386.

Bank, W. J. (1980). Thallium. In *Experimental and Clinical Neurotoxicology,* Spencer, P. S., and Schaumburg, H. H. (eds.). Williams and Wilkins, Baltimore, pp. 570–577.

Bank, W. J., Pleasure, D. E., Susuki, J., Nigro, M., and Katz, R. (1972). Thallium poisoning. *Arch. Neurol.,* **26,** 456–464.

Barbier, F. (1974). Treatment of thallium poisoning. *Lancet,* **11,** 965.

Barclay, R. K., Peacok, W. C., and Karnovsky, D. A. (1953). Distribution and excretion of radioactive thallium in the chick embryo, rat, and man. *J. Pharm. Pharmacol.,* **107,** 178–186.

Barnes, M. P., Murray, K., and Tilley, P. J. B. (1984). Neurological deficit more than thirty years after chronic thallium intoxication (letter). *Lancet,* **1,** 184.

Barroso-Moguel, R., Galban Arzate, S., Villeda-Hernández, J., Mendez-Armenta, M., Alcaraz-Zubeldia, M. M., and Rios C. (1994a). Antitumoral effect of thallium against *N*-ethyl-*N*-nitroso-urea-induced brain tumors. *Proc. West. Pharmacol. Soc.,* **37,** 27–28.

Barroso-Moguel, R., Villeda-Hernández, J., Mendez-Armenta, M., Rios, C., and Monroy-Noyola, A. (1994b). Combined D-penicillamine and Prussian blue as antidotal treatment against thallo-toxicosis in rats: Evaluation of cerebellar lesions. *Toxicology,* **89,** 15–24.

Baselt, R. C., and Cravey, R. H. (1995). *Disposition of Toxic Drugs and Chemicals in Man,* 4th ed. Chemical Toxicology Institute, Foster City, CA, pp. 717–719.

Bergquist, J. E., Edström, A., and Hansson, P. A. (1981). Bidirectional axonal transport of thallium in rat sciatic nerve. *Neurology,* **31,** 612–616.

Bertini, I., Luchinat, C., and Mesori, L. (1983). Tl as an NMR probe for the investigation of transferrin. *J. Am. Chem. Soc.,* **105,** 1347–1350.

Bertrán-Capella, A., Hernández-Gutierrez, F., and Corbella, J. (1972). *Intoxication au Thallium.* Masson and Cia, Lyon, p. 219.

Britten, J. S., and Blank, M. (1968). Thallium activation of the (Na⁺-K⁺)-activated ATPase of rabbit kidney. *Biochim. Biophys. Acta*, **159**, 160–166.

Brockhaus, A., Dolgner, R., Ewers, U., Krömer, U., Soddeman, H., and Wiegand, H. (1981). Intake and health effects of thallium among a population living in the vicinity of a cement plant emitting thallium containing dust. *Int. Arch. Occ. Environ. Hlth.*, **48**, 375–389.

Buschke, A., and Peiser, B. (1923). Epithelial proliferation in the fore-stomach of rats by the experimental action of thallium. *Z Krebsforsch* (Berlin), **21**, 11–18.

Canther, H. E. (1974). Biochemistry of selenium. In *Selenium*, Zingaro, R. A., and Cooper, W. C. (eds.). Van Nostrand Reinhold, New York, pp. 564–614.

Careaga-Olivares, J., and González-Ramírez, D. (1995). Penicillamine produces changes in the acute blood elimination and tissue accumulation of thallium. *Arch. Med. Res.*, **26**, 427–430.

Cavanagh, J. B. (1979a). Metallic toxicity and the nervous system. *Recent Adv. Neuropathol.*, **1**, 247–275.

Cavanagh, J. B. (1979b). The dying back process. *Arch. Pathol. Lab. Med.*, **103**, 659—669.

Cavanagh, J. B. (1991). What have we learnt from Graham Frederick Young? Reflections on the mechanism of thallium neurotoxicity. *Neuropathol. Appl. Neurobiol.*, **17**, 3–9.

Cavanagh, J. B., and Gregson, L. (1978). Some effects of thallium salt on the proliferation of hair follicle cells. *J. Pathol.*, **125**, 179.

Cavanagh, J. B., Fuller, N. H., Johnson, H. R. M., and Rudge, P. (1974). The effects of thallium salts, with particular reference to the nervous system changes. *Quart. Med. J.*, **43**, 293–319.

CDC (1987). Thallium poisoning—an epidemic of false positives, Georgetown, Guyana. *M.M.W.R.*, **36**, 481–488.

Chandler, H. A., and Scott, M. (1986). A review of thallium toxicology. *J. Roy. Nav. Med. Ser.*, **72**, 75–79.

Chandler, H. A., Archbold, G. P. R., Gibson, J. M., O'Callagham, P., Marks, J. N., and Petthbridge, R. J. (1990). Excretion of a toxic dose of thallium. *Clin. Chem.*, **36**, 1506–1509.

Crafoord, B., Bondesson, I., Clemedson, C., Löving, T., Sandler, H., and Ekwall, B. (1996). Thallium. In *MEIC Monographs on Time-related, High Survived and Lethal Blood Concentrations of Chemicals from Acute, Human Poisonings*, Ekwall, B. (ed.). CTLU, Uppsala.

Davis, L. E., Standefer, J. C., Kornfeld, M., Abercrombie, D. M., and Butler, C. (1981). Acute thallium poisoning: Toxicological and morphological studies of the nervous system. *Ann. Neurol.*, **10**, 38–44.

de Groot, G., van Heijst, A. N. P., van Kesteren, R. G., and Maes, R. A. A. (1985). An evaluation of the efficacy of charcoal haemoperfusion in the treatment of three cases of acute thallium poisoning. *Arch. Toxicol.*, **57**, 61–6.

Desenclos, J. C., Wilder, M. H., Coppenger, G. W., Sherin, K, Tiller, R., and VanHook, R. M. (1992). Thallium poisoning: An outbreak in Florida, 1988. *Southern Med. J.*, **85**, 1203–1206.

Dolgner, R., Brockhaus, A., Ewers, U., Wiegand, H., Majewski, F., Soddemann, H. (1983). Repeated surveillance of exposure to thallium in a population living in the vicinity of a cement plant emitting dust containing thallium. *Int. Arch. Occ. Environ. Hlth.*, **52**, 79–94.

Douglas, K. T., Bunni, M. A., and Baindur, S. R. (1990). Thallium in biochemistry. *Int. J. Biochem.*, **22**, 429–438.

Dumitru D., and Kalantri, A. (1990). Electrophysiologic investigation of thallium poisoning. *Muscle Nerve*, **13**, 433–437.

EPA (1992). 40 CFR Part 355, Appendices A and B; Emergency Planning and Notification; The List of Extremely Hazardous Substances and their Threshold Planning Quantities Environmental Protection Agency, Washington, DC.

Feldman, J., and Levisohn, D. R. (1993). Acute alopecia: Clue to thallium toxicity. *Pediatr. Dermatol.*, **10**, 29–31.

Formigli, L., Scelsi, R., Poggi, P., Gregotti, C., Di Nucci, A., Sabbioni, F., Gottardi, L., and Manzo, L. (1986). Thallium-induced testicular toxicity in the rat. *Environ. Res.,* **40,** 531–539.

Fowler, B. A., Yamauchi, H., Conner, E. A., and Akkerman, M. (1993). Cancer risks for humans from exposure to the semiconductor metals. *Scand. J. Work. Environ. Hlth.,* **1,** 101–103.

Galván-Arzate, S., and Rios, C. (1994). Thallium distribution in organs and brain regions of developing rats. *Toxicology,* **90,** 63–69.

Gehring, P., and Hammond, T. (1967). The inter-relationship between thallium and potassium in animals. *Pharmacol. Exp. Ther.,* **155,** 187–201.

Gettler, A., and Weiss, L. (1943). Thallium poisoning. III. Clinical toxicology of thallium. *Am. J. Clin. Pathol.,* **13,** 422–429.

Gibson, J. E., and Becker, B. A. (1970). Placental transfer, embryotoxicity, and teratogenicity of thallium sulphate in normal and potassium-deficient rats. *Toxicol. Appl. Pharmacol.,* **16,** 120–132.

Grant, W. M. (1986). *Toxicology of the Eye,* 3rd ed. Charles C. Thomas, Springfield, IL.

Grunfeld, O., and Hinostroza, G. (1964). Thallium poisoning. *Arch. Intern. Med.,* **114,** 132–134.

Hasan, M., and Ali, S. F. (1981). Effects of thallium, nickel, and cobalt administration on the lipid peroxidation in different regions of the rat brain. *Toxicol. Appl. Pharmacol.,* **57,** 8–13.

Hasan, M., Chandra, S. V., and Bajpai, V. K. (1977a). Electron microscopic effects of thallium poisoning on the rat hypothalamus and hippocampus: Biochemical changes in the cerebrum. *Brain Res. Bull.,* **2,** 255–261.

Hasan, M., Chandra, S. V., Dua, P. R., Raghubir, R., and Ali, S. F. (1977b). Biochemical and electrophysiological effects of thallium poisoning on the rat corpus striatum. *Toxicol. Appl. Pharmacol.,* **41,** 353–359.

Hennekes, R. (1983). Netzhautfunktions schodigung bei akuter Thalliumintoxikation. *Klin. Mbl. Augenheilk.,* **182,** 334–336.

Herrero, F., Fernandez, E., Gomez, J., Pretel, L., Cañizares, F., Frias, J., and Escribano, J. B. (1995). Thallium poisoning presenting with abdominal colic, paresthesia, and irritability. *Clin. Toxicol.,* **33,** 261–264.

Heydlauf, H. (1969). Ferric-cyanoferrate (II): An effective antidote in thallium poisoning. *Eur. J. Pharmacol.,* **6,** 340–344.

Hologgitas, J., Ullicci, P., Driscoll, J., Gauerholz, J., and Martin, H. (1980). Thallium elimination kinetics in acute thallotoxicosis. *J. Analytical. Toxicol.,* **4,** 68–73.

Huber, F., and Kirchmann, H. (1978). Biomethylation of Tl(I) compounds. *Inorg. Chem. Acta,* **29,** L249–L250.

Hurlbut, K. M., Dart, R. C., Sullivan, J. B., and Campbell, D. S. (1990). Use of DMSA in severe thallium toxicity. *Vet. Hum. Toxicol.,* **32**(4), 363.

Hutlin, T., and Naslund, P. H. (1974). Effects of thallium(I) on the structure and function of mammalian ribosomes. *Chem-Biol. Interact.,* **5,** 315–328.

Insley, B. M., Grufferman, S., and Ayliffe H. E. (1986). Thallium poisoning in cocaine abusers. *Am. J. Emerg. Med.,* **4,** 545–548.

Inturrisi, C. E. (1969). Thallium-induced dephosphorylation of a phosphorylated intermediate of the (sodium + thallium-activated) ATPase. *Biochim. Biophys. Acta,* **178,** 630–633.

Kamerbeek, H. H., Rauws, A. G., ten Ham, M., and van Heijst, A. N. P. (1971). Prussian Blue in therapy of thallotoxicosis. *Acta. Med. Scand.,* **189,** 321–324.

Kamil'dzhanov, A. K. (1993). Experimental substantiation of maximum permissible concentration of thallium carbonate in environmental air. *Gig. Sanit.,* **5,** 8–10.

Kaye, J. F. (1971). Thallium(I) activation of pyruvate kinase. *Arch. Biochem. Biophys.,* **143,** 232–239.

Kazantzis, G. (1986). Thallium. In *Handbook on the Toxicology of Metals,* Friber, L., Nordberg, G. F., and Vouk, V. (eds.). Elsevier Science, Amsterdam, pp. 549–567.

Kennedy, P., and Cavanagh, J. B. (1977). Sensory neuropathy produced in the cat with thallous acetate. *Acta Neuropathol.*, **39**, 81–88.

van Kesteren, R. G., Rauws, A. G., de Groot, G., and van Heijst A. N. P. (1980). Thallium intoxication. An evaluation of therapy. *Intensive. Med.*, **17**, 293–297.

Komulainen, H., and Bondy, S. C. (1988). Increased free intracellular Ca^{2+} by toxic agents: an index of potential neurotoxicity? *Trends Pharmacol. Sci.*, **9**, 154–156.

Koshy, K. M., and Lovejoy, F. H. (1981). Thallium ingestion with survival ineffectiveness of peritoneal dialysis and potassium chloride diuresis. *Clin. Toxicol.*, **18**, 521–525.

Kravzov, J., Rios, C., Altagracia, M., Monroy-Noyola A., and Loópez, F. (1993). Relationship between physicochemical properties of Prussian blue and its efficacy as antidote against thallium poisoning. *J. Appl. Toxicol.*, **13**, 213–216.

Kuhn, R., Rudy, H., and Wagner-Jauregg, T. (1933). Uber lactoflavin (Vit. B_2). *Berichte Deutschen Chemischen Gessellschaft*, **66**, 1950–1956.

Labauge, R., Pages, M., Tourniaire, D., Favier, J. P., and Blard, J. M. (1991). Recurrent peripheral neuropathy caused by thallium poisoning. *Rev. Neurol.*, **147**, 317–319.

Lameijer, A., and van Zwieten, P. A. (1977). Kinetic behavior of thallium in the rat. Accelerated elimination owing to treatment with potent diuretic agents. *Arch. Toxicol.*, **37**, 265–273.

Lehmann, P. A., and Favari, L. F. (1984). Parameters for the adsorption of thallium ions by activated charcoal and Prussian blue. *Clin. Toxicol.*, **22**, 331–339.

Lehmann, P. A., and Favari, L. F. (1985). Acute thallium intoxication: Kinetic study of the relative efficacy of several antidotal treatments in rats. *Arch. Toxicol.*, **57**, 56–60.

Lie, R., Thomas, R. G., and Scott, J. K. (1960). The distribution and excretion of thallium-240 in the rat, with suggested MCP's and a bioassay procedure. *Hlth Phys.*, **2**, 334–340.

Lovejoy, F. H. (1982). Thallium. *Clin. Toxicol. Rev.*, **4**, 1–2.

Luckit, J., Mir, N., Hargraves, M., Costello, C., and Gazzard, B. (1990). Thrombocytopenia associated with thallium poisoning. *Human Exp. Toxicol.* **9**, 47–48.

Lund, A. (1956a). Distribution of thallium in the organs and its elimination. *Acta Pharmacol. Toxicol.*, **12**, 251–259.

Lund, A. (1956b). The effect of various substances on the excretion and toxicity of thallium in the rat. *Acta Pharmacol. Toxicol.*, **12**, 260–268.

van Maarseveen, D. A. (1962). Een geval van thallium-intoxicatie tijdens de zwangerschap. *Ned. Tijdschr. Geneeskd.*, **106**, 1765–1766.

Manzo, L., and Sabbioni, E. (1988). Thallium. In *Handbook on Toxicity of Inorganic Compounds.* Seiler, H. G., Sigel, A., and Sigel, H. (eds). Marcel Dekker, New York, pp. 677–688.

Marcus, R. L. (1985). Investigation of a working population exposed to thallium. *J. Soc. Occup. Med.*, **35**, 4–9.

Meggs, W. J., Hofman, S., Shih, R. D., Weisman, R. S., and Goldfrank, L. R. (1994). Thallium poisoning from maliciously contaminated food. *Clin. Toxicol.*, **32**, 723–730.

Melnick, R. L., Monti, L. G., and Motzkin, S. M. (1976). Uncoupling of mitochondrial oxidative phosphorylation by thallium. *Biochem. Biophy. Res. Commun.*, **69**, 68–73.

Minoia, C., Sabbioni, E., Apostoli, P., Pietra, R., Pozzoli, L., Gallorini, M., Nicolau, G., Alessio, L., and Capodaglio, E. (1990). Trace elements reference values from inhabitants of the European Community. *Sci. Tot. Environ.*, **95**, 89–110.

Moeschlin, S. (1980). Thallium poisoning. *Clin. Toxicol.*, **17**, 133–146.

Moore, D., House, I., and Dixon, A. (1993). Thallium poisoning. *Br. Med. J.*, **306**, 1527–1529.

Mulkey, J. P., and Oehme, F. W. (1993). A review of thallium toxicity. *Vet. Hum. Toxicol.*, **35**, 445–453.

Mullins, L. J., and Moore, R. D. (1960). The movement of thallium ions in muscle. *J. Gen. Physiol.*, **43**, 759–773.

Munch. J. C., (1934). Human thallotoxicosis. *J. Am. Med. Assoc.* **102,** 1929–1934.

Munch, J. C., Ginsberg, H. M., and Nixon, C. E. (1933). The 1932 thallotoxicosis outbreak in California. *JAMA,* **100,** 1315–1319.

Niehues, R., Horstkotte, D., Klein, R. M., Kuhl, U., Kutkuhn, B., Hort, W., Iffland, R., and Strauer, B. E. (1995). Wiederholte ingestion potentiell letaler thalliummengen in suizidaler absicht. *Dtsh. Med. Wochenschr.,* **120,** 403–408.

NIOSH (1992). *Registry of Toxic Effects of Chemical Substances.* National Institute for Occupational Safety and Health, Cincinnati, OH (CD-ROM Version), Englewood, CO.

Nogué, S., Mas, A., Parés, A., Nadal, P., Bertrán, A., Millá, J., Carrera, M., To., J., Pazos, M. R., and Corbella, J. (1982–83). Acute thallium poisoning: An evaluation of different forms of treatment. *J. Toxicol. Clin. Toxicol.,* **19,** 1015–1021.

Omura, Y., Lorberboym, M., and Beckman, S. (1995). Radiation injury and mercury deposits in internal organs as a result of thallium-201 chloride intravenous injection for SPECT imaging: Additional biochemical information obtained in the images of organs from SPECT or PET scans; & potential injury due to radiation exposure during long distance flights. *Acupunct. Electrother. Res.,* **20,** 133–148.

Papp, J. P., Gay, P. C., Dodson, V. N., and Pollard, H. M. (1969). Potassium chloride in the treatment of thallotoxicosis. *Ann. Intern. Med.,* **71,** 119–123.

Paulson, G., Vergara, G., Young, J., and Bird, M. (1972). Thallium intoxication treated with dithizone and hemodialysis. *Arch. Intern. Med.,* **129,** 100–103.

Pedersen, R. S., Olesen, A. S., Freund, L. G., Solgaard, P., and Larsen, E. (1978). Thallium intoxication treated with long-term hemodialysis, forced diuresis and Prussian blue. *Acta Med. Scand.,* **204,** 429–432.

Petersohn, K. L. (1960). Thallium-Vergiftungen in der Schwangerschaft. *Arch. Toxicol.,* **18,** 160–164.

Piazolo, P., Franz, H. E., Brech, W., Walb, D., and Wilk, G. (1971). Behandlung der Thalliumvergiftung mit der Hamodialyse. *Dtch. Med. Wschl.,* **96,** 1217–1222.

Prick, J. J. (1979). Thallium poisoning. In *Handbook of Clinical Neurology,* Vol. 36, Vinken, P. J., and Bruyn, G. W. (eds.). Elsevier, North Holland, New York, pp. 239–278.

Prick, J. J. G., Sillevis Smitt, W. G., and Muller, L. (1955). *Thallium Poisoning.* Elsevier, Amsterdam.

Questel, F., Dugarin, J., and Dally, S. (1996). Thallium-contaminated heroin (letter). *Ann. Intern. Med.,* **124,** 616.

Rangel-Guerra, R., Martínez, H. R., and Villareal, H. J. (1990). Thallium intoxication. Review of 50 cases (in Spanish). *Gac. Med. Mex.,* **126,** 487–494.

Rasmussen, O. V. (1981). Thallium poisoning; an aspect of human cruelty. *Lancet,* **1,** 1164.

Rauws, A. G. (1974). Thallium pharmacokinetics and its modification by Prussian Blue. *Naunyn-Schmiedeberg's Arch. Pharmacol.,* **285,** 295–306.

Reed, D., Crawley, J., Faro, S., Pieper, S., and Kurland, L. (1963). Thallotoxicosis, acute manifestations and sequelae. *J. Am. Med. Assoc.,* **183,** 516–522.

Repetto, M. and López-Artíguez, M. (1977). A multiple case of thallium poisoning. Report 3928, National Institute of Toxicology, Seville, Spain.

Repetto, G., and Sanz, P. (1993). Neutral red uptake, cellular growth and lysosomal function: *in vitro* effects of 24 metals. *Alt. Lab. Anim.,* **21,** 501–507.

Repetto, G., Sanz, P., and Repetto, M. (1994). *In vitro* effects of thallium on mouse neuroblastoma cells. *Toxicol. in Vitro,* **8,** 609–611.

Richelmi, P., Bono, F., Guarda, L., Ferrini, B., and Manzo, L. (1980). Salivary levels of thallium in acute human poisoning. *Arch. Toxicol.,* **43,** 321–325.

Richeson, E. M. (1958). Industrial thallium intoxication. *Ind. Med. Surg.,* **27,** 607–619.

Rios, C., and Monroy-Noyola, A. (1992). D-penicillamine and Prussian blue as antidotes against thallium intoxication in rats. *Toxicology,* **74**, 69–76.

Rios, C., Galván-Arzate, S., and Tapia, R. (1989). Brain regional thallium distribution in rats acutely intoxicated with $Tl_2 SO_4$. *Arch. Toxicol.,* **63**, 34–37.

Roby, D. S., Fein, A. M., Bennett, R. H., Morgan, I. S., Zatuchini, J., and Lippmann, M. L. (1984). Cardiopulmonary effects of acute thallium poisoning. *Chest,* **85**, 236–240.

Sabbioni, E., Gregotii, C., Edel, J., Marafante, E., Di Nucci, A., and Manzo, L. (1982). Organ/ tissue disposition of thallium in pregnant rats. *Arch. Toxicol.* (suppl.), **5**, 225–230.

Sabbioni, E., Ceotz, L., and Bignoli, G. (1984). Health and environmental implications of trace metals released from coal-fired power plants: An assessment study of the situation in the European Community. *Sci. Total. Environ.,* **40**, 141–154.

Saddique, A., and Peterson, C. D. (1983). Thallium poisoning. A review. *Vet. Hum. Toxicol.,* **25**, 16–22.

Sax N. I. (1989). *Dangerous Properties of Industrial Materials,* Van Nostrand Reinhold, New York.

Schaller, K. H., Manke, G., Raithel, H. J., Buehlrneyer, G., Schmidt, S. I. and Valentin, H. (1980). Investigations of thallium-exposed workers in cement factories. *Int. Arch. Occup. Environ. Hlth.,* **47**, 223–231.

Schaumburg, H. H., and Berger, A. (1992). Alopecia and sensory polyneuropathy from thallium in a Chinese herbal medication (letter). *JAMA,* **268**, 2430–2431.

Schmidbauer, H., and Klingler, D. (1979). Chronische Thalliumvergiftung. *Wien. Med. Wschr.,* **129**, 334–336.

Schoental, R., and Cavanagh, J. B. (1977). Mechanism involved in the "dying back" process—an hypothesis involving cofactor. *Neuropathol. Appl. Neurobiol.,* **3**, 145.

Shabalina, L. P., and Spiridonova, U. S. (1979). Thallium as an industrial poison. Review of literature. *J. Hyg. Epidemiol. Microbiol. Immunol.,* **231**, 247–256.

Shamshinova, A. M., Ivanina, T. A., Yakovlev, A. A., Shabalina, L. P., and Spiridonova, V. S. (1990). Electroretinography in the diagnosis of thallium intoxication. *J. Hyg. Epidemiol. Microbiol. Immunol.,* **34**, 113–121.

Sikkel, A., Posthumes Meyes, F. E., and Zondag, H. A. (1959). Een geval van thallium vergiftiging in de zwangerscha. *Ned Tijdschr Geneeskd.* **59**, 386–394.

Singh, N. P., Bodgen, J. D., and Joselow, M. M. (1975). Distribution of thallium and lead in children's blood. *Arch Environ Hlth.,* **30**, 557–558.

Smith, D. H., and Doherty, R. A. (1964). Thallotoxicosis. Report of three cases in Massachusetts. *Pediatrics,* **34**, 480–490.

Smith, I. C., and Carson, B. L. (1977). Thallium. In *Trace Metals in the Environment,* Vol. I. Ann Arbor Science, Ann Arbor, MI, p. 45.

Spencer, P. S., Peterson, E. R., Madrid, M. A., and Raine, C. S. (1973). Effects of thallium salts on neuronal mitochondria in organotypic cord-ganglia-muscle combination cultures. *J. Cell. Biol.,* **58**, 79–95.

Stevens, W. J. (1978). Thallium intoxication caused by a homeopathic preparation. *Toxicol. Eur. Res.,* **1**, 317–320.

Stevens, W. J., and Barbier, F. (1976). Thallium intoxicatie gedurende zwangerschap. *Act. Clin. Belg.,* **31**, 188–193.

Stevens, W., van Peteghem, C., Heyndrickx, A., and Barbier, F. (1974). Eleven cases of thallium intoxication treated with Prussian Blue. *Int. J. Clin. Pharmacol.,* **10**, 1–22.

Stokinger, H. E. (1981). The Metals. In *Patty's Industrial Hygiene and Toxicology,* Clayton, G. T., and Clayton F. E., (eds.), Wiley, New York, pp. 1493–2060.

Swank R. L., Porter R. R., and Yeomans A. (1941). The production and study of cardiac failure in thiamine deficient dogs. *Am. Heart J.,* **22**, 154–168.

Symmonds, W. J. C. (1953). Alopecia, optic atrophy and peripheral neuritis of probably toxic origin. *Lancet,* **11,** 1338–1339.

Tabandeh, H., Crowston, J. G., and Thompson, G. M. (1994). Ophthalmologic features of thallium poisoning. *Am. J. Ophthalmol.,* **117,** 243–245.

Talas, A., and Wellhöner, H. H. (1983). Dose-dependency of Tl^+ kinetics as studied in rabbits. *Arch. Toxicol.* **53,** 9–16.

Talas, A., Pretschner, D. P., and Wellhöner, H. (1983). Pharmacokinetic parameters for thallium(I) ions in man. *Arch. Toxicol.* **53,** 1–7.

Thompson, C., Dent, J., and Saxby, P. (1988). Effects of thallium poisoning on intellectual function. *Br. J. Psychiatry,* **153,** 396–399.

Thyresson, N. (1951). Experimental investigation on thallium poisoning. Effect of thallium on the growth and differentiation of hairs and epidermis, studied in young rats and in tissue cultures of embryonic rat skin. *Acta. Derm. Venereol.,* **31,** 133–146.

Truhaut, R. (1958). L'intoxication par le thallium. Revue générale et présentation des résultats personnels sur le double plan analytique et biologique. *Ann. Med. Lég.,* **38,** 189–239.

van der Merwe, C. F. (1972). The treatment of thallium poisoning. *S. Afr. Med. J.,* **46,** 960–961.

Venugopal, B., and Luckey, T. D. (1978). *Metal Toxicity in Mammals,* Vol. 2. *Chemical Toxicity of Metals and Metalloids.* Plenum, New York, p. 307.

Vegauwe, P. L., Knockaert, D. C., and Van Tittelboom, T. J. (1990). Near fatal subacute thallium poisoning necessitating prolonged mechanical ventilation. *Am. J. Emerg. Med.,* **8,** 548–550.

Vrij, A. A., Cremers, H. M., and Lustermans, F. A. (1995). Successful recovery of a patient with thallium poisoning. *Neth. J. Med.,* **47,** 121–126.

Wainwright, A. P., Kox, W. J., House, I. M., Henry, J. A., Heaton, R., and Seed, W. A. (1988). Clinical features and therapy of acute thallium poisoning. *Quart. J. Med.,* **69,** 939–944.

Wakid, N. W., and Cortas, N. K. (1984). Chemical and atomic absorption methods for thallium in urine compared (letter). *Clin. Chem.,* **30,** 587–588.

Weinig, E., and Zink, P. (1967). Quantitative mass spectrometric determination of the normal thallium content in the human body. *Arch. Toxicol.,* **22,** 275–281.

Williams, R. J. P. (1970). The biochemistry of sodium, potassium, magnesium, and calcium. *Q. Rev. Chem. Soc.,* **24,** 331–365.

WHO (1973). Safe use of pesticides: 20th report of WHO expert committee on insecticides. World Health Organization Technical Reference Services, 513, 40.

Ziskoven, R., Achenbach, C., Ba, U., and Schulten, H. R. (1980). Quantitative investigations on the diaplacental transfer of thallium by field desorption mass spectrometry. *Z. Naturf.,* **35c,** 902–906.

10

REPRODUCTIVE AND DEVELOPMENTAL TOXICITY OF THALLIUM

Cesarina Gregotti

Institute of Pharmacology, Medical School, University of Pavia, Piazza Botta 10, 27100 Pavia, Italy

Elaine M. Faustman

Department of Environmental Health, University of Washington, 422 Roosevelt Way NE, 100, Seattle, Washington 98105-6099

Thallium in the Environment, Edited by Jerome O. Nriagu.
ISBN 0-471-17755-5. © 1998 John Wiley & Sons, Inc.

10.1. INTRODUCTION

Thallium, which is a natural constituent of Earth's crust, is a rare but ubiquitous element, and occurs in low amounts in almost all living organisms. In the past, thallium compounds were used medicinally for the treatment of syphilis, gonorrhea, gout, dysentery, night sweats in tuberculosis, and as a cosmetic depilatory to remove excessive growth of hair (Prick et al., 1955). Also, this metal in the form of sulfate was employed as a rodenticide and as an insecticide. All of these applications of thallium caused poisoning of human beings. Severe intoxications and fatalities were reported following accidental or deliberate intake (Cavanagh et al., 1974), and toxic effects have been well documented in a wide variety of biological systems, ranging from mammals to plants and microorganisms (Sabbioni and Manzo, 1980; Douglas et al., 1990). For these reasons the medicinal use of thallium was abandoned, although its use for pest control is still allowed in many countries because of its high efficacy. More recently, thallium toxicity has caused concern due, to its possible release in the environment following industrial use (Smith and Carson, 1977; Zitko, 1975). While chronic occupational intoxication is infrequent and the industrial use of thallium compounds is rare, adverse health effects of long-term environmental exposure to trace amounts of thallium from various anthropogenic sources (coal-burning power plants, smelters, refineries, iron and steel industries, and burning fuel) are under consideration. In particular, although studies of thallium toxicity have focused primarily on the nervous system. (Cavanagh et al., 1974; Manzo et al., 1983), skin (Cavanagh and Gregson, 1978), and cardiovascular tract (Lameijer and Van Zwieten, 1976), the adverse effects of this chemical on the reproductive system is an area of growing concern as the assessment of potential fertility disorders has become increasingly important in both experimental toxicology (Gunn and Gould, 1970; Dixon and Hall, 1982) and epidemiological studies.

10.2. MUTAGENICITY

10.2.1. Bacteria and Amphybia

Thallium at high levels has been shown to inhibit the reproductive rate of bacteria. Concentrations ranging between 0.8 and 8 ppm of thallium inhibited nitrification by *Nitrobacter agilis,* and the growth of *Pseudomonas aeruginosa* was prevented by a thallium acetate solution of 120 mg/L (Tandon and Mishra, 1968). However, 39 m*M* TI did not inhibit growth of *Staphylococcus aureus,* and *Mycoplasmataceae* were more resistant than *Acholeplasmaceae* (Kunze, 1972). Thallium acetate retarded both mitosis and meiosis of larvae and nymphs. Moreover, it was shown to block differentiation and reproduction of protozoa and metamorphosis of amphybia (Sabbioni and Manzo, 1980).

10.2.2. Mammalian Cells

The mutagenic effect of thallium was demonstrated in cell cultures inoculated with variola vaccine and then treated with thallium carbonate $10^{-4}\,M$ for 24 h. Also monolayer cultures of embryo fibroblasts treated with the same protocol (time, dose, and salt) caused chromosome aberrations and increased the frequency of DNA breaks followed by almost complete repair after a recovery period of 24 h (Zashukina et al., 1980). Further studies revealed similar mutagenic activity in cultures of embryo cells of mice and rats (Zashukina et al., 1983). Investigation of the dominant mutation frequency in intact white rat females mated with males chronically treated with thallium carbonate (5×10^{-5} and 5×10^{-4} mg/kg body weight) showed a tendency toward enhancement of general embryotoxicy with a wider number of resorptions and post-implantation deaths at a concentration of 5×10^{-6} mg/kg (Zashukina et al., 1983).

In human lymphocytes cultured in vitro with PHA (Phyto-Haemo-Agglutinine), a mitosis stimulant, thallium acetate reduced the uptake of tritiated thymidine and decreased the mitotic index. Inhibition of DNA synthesis and mitosis was found to be linearly related to thallium dose (EC_{50} 2γ/mL) on a semilogarithmic scale (Governa et al., 1969).

Because of the toxicity to both mammalian germ cells in vivo and somatic cells in vitro of ^{201}Tl used in nuclear medicine (Rao et al., 1983; Kassis et al., 1983), Kelsey et al. (1991) investigated a possible mutagenic effect of in vivo exposure to low levels of this ionizing radiation in 24 patients. The hypoxanthine guanine phosphoribosyl transferase mutant fraction (MF) and chromosome aberration (CA) frequency were examined before and 24 h after treatment. No significant elevation of MF and CA frequencies in lymphocytes of exposed individuals was found.

10.3. CARCINOGENICITY

Female mice treated orally or cutaneously with high doses of thallium showed a degenerative process in the genital tract similar to the one found in castrated animals or after uterine denervation. The diagnoses were papilloma, precancerous lesions, and cancer. The control mice did not develop cancer. These alterations were not observed in the control group of mice that did not develop cancer spontaneously (Champy et al., 1958). Hart and Adamson (1971), after intraperitoneal injection of 3.5 mg/kg thallium chloride in rats carrying ascitic Walker 256 carcinoma, observed an increase of median survival time and of the number of long-term survivors.

10.4. TERATOGENICITY

10.4.1. In Vitro

Rat embryonic cultures exposed to thallium (3–100 μg Tl/mL) for 48 h showed a dose–dependent development of retardation. Complete growth inhibition was observed at the highest dose, and central nervous system (CNS) development was drastically impaired. At less than 10 mg Tl/mL normal development of rat embryos was macroscopically observed, whereas histological analysis revealed cytotoxic effects (Anschuetz et al., 1981).

10.4.2. In Vivo

10.4.2.1. Laboratory Animal Data

In 1950, Karnofky et al. showed that Tl_2SO_4, 0.4–0.7 mg/egg administered onto the chorioallantoic membrane on day 8 or by yolk sac injection on day 4 of incubation, produced 90–100% incidence of achondroplasia in chicken embryos.

Hall (1976) demonstrated that there was a "sensitive" period during early embryonic development (day 5–8.5) for induction of thallium malformations. No embryos survived after thallium injection of eggs on days 0–4. Eggs injected at or after 9 days of incubation had normal body weights and no signs of tibial bending. Thallium administered to eggs in between this time resulted in a significant reduction of embryo body weight compared to controls and reduction in the growth of the tibias where histological examination revealed chondrocytic necrosis.

In order to elucidate whether thallium acts directly on the skeleton, Hall (1976) grafted tibias of thallium-treated and control embryos to the chorioallantonic membranes of treated and control host embryos.

According to Hall (1976), direct and rapid incorporation of thallium into the tibia may be due to an enzymatic block, as first reported by Landauer

(1960), and thallium may also act as a substitute of potassium in activating ATPase in other potassium requiring enzymes (Cavieres and Ellory, 1974).

Gibson and Becker (1970) administered thallium sulfate to pregnant rats on days 12, 13, and 14 in order to compare the teratogenic effects of thallium in rodents with those produced during treatment in this same period of chick embryo organogenesis. Since the chick embryos were generally more sensitive to the toxic effects of thallium in the earlier stage of development, the rats were also intoxicated during that period of gestation.

Thallium administered to rats during early (2.5 mg/kg) or late (10 mg/kg) embryogenesis reduced fetal body weight but did not have a significant effect on embryo or fetal resorptions. In a few cases treatment during late gestation induced teratogenic effects. These effects were limited to missing vertebral bones and hydronephrosis, which was probably related to the concentrations of thallium in the kidney.

A teratogenic study previously done by the same authors (Gibson et al., 1967) in pregnant females of different species (mice, rabbits, and rats) indicated that the administration of thallium sulfate produced only slight toxic or teratogenic effects on developing embryos or fetuses even when the mothers were treated with toxic amounts of thallium before and throughout the entire gestation period. Massive teratogenic effects in mice treated with thallium sulfate were described by Achenbach et al. (1979a;b), but their results failed to be confirmed by other investigators (Claussen et al., 1981).

Whole-body autoradiography was used to study thallium uptake and retention in mice during gestation. Olsen and Jonsen (1982) injected ^{204}Tl intraperitoneally to pregnant mice at 5–16 days of gestation. The results of autoradiography indicated that thallium crossed the placental barrier throughout gestation. During early gestation thallium was retained by the visceral yolk sac and during late gestation even by the chorioallantoic placenta and amnion. Moreover, large amounts of thallium crossed the placental barrier of rats at 13 days of gestation, reaching appreciable levels in fetal tissues as shown in Table 1.

10.4.2.2. Human Data

As a result of attempted suicide, attempted abortion, or accidental ingestion, many cases of thallium toxicosis have been reported in pregnant subjects (Erbsloh, 1960; Johnson, 1960; Petersohn, 1960; Moeschlin, 1980). The quantity of thallium sulfate swallowed, at different stages of pregnancy, ranged from 0.35 to 1.2 g as thallium. In most cases thallium was present in the mother's urine and feces before and after labor, and in one case it was also detected in the placenta, in amniotic fluid, and in milk (Moeschlin, 1980). In addition, human fetuses appear to suffer from transplacental thallium toxicosis as evidenced by low birth weight and alopecia. Alopecia in newborns is typical of thallium skin and nail growth disturbances seen in mammals (see Section 10.4.2.1.). However, no evidence of congenital abnormalities have been found. In all cases the mother appeared to be seriously affected whereas in the fetus

Table 1   Thallium Concentrations in Selected Tissues of Pregnant Rats and Fetus 72 h after Treatment (25 mg/kg Tl_2SO_4 p.o.)

Tissue	Thallium ($\mu g/g$)
Pregnant rats	
Liver	26.3 ± 2.0
Brain	21.8 ± 1.5
Blood	3.4 ± 0.8
Small intestine	65.0 ± 6.3
Fetuses	
Liver	13.1 ± 1.8
Brain	14.1 ± 1.1
Whole fetuses	33.3 ± 1.7
Placenta	22.0 ± 1.8
Amniotic fluid	1.4 ± 0.2

Source: Di Nucci et al., 1979.

the severity of thallium-intoxication signs depended on time of occurrence during pregnancy and dose ingested.

10.4.2.3. Environmental Exposure

In Lengerich, West Germany (about 20,000 inhabitants), damage to plants and domestic animals had been observed in the vicinity of cement plants for several years, but only in August, 1979, was it understood that this damage was caused by the cement plant atmospheric dust fallout containing thallium (Prinz et al., 1979). Attempted analyses of the production process showed that the emission of thallium was caused by a residue of pyrite roasting, which was added, as an additive containing ferric oxide, to the crude powered limestone for the production of a special type of cement. Chemical analyses revealed that the thallium content of this additive was about 400 ppm, and the total emission of thallium has been estimated to be about 5 kg/24 h. Starting from August, 1979, the emission of thallium was reduced more than hundredfold with the introduction of changes in technology and the employment of additives with not more than 2 ppm of thallium.

From 1979 to 1981, several medical surveys were carried out among the population living in the vicinity of the plant to assess the degree of thallium exposure and to investigate health disorders that might be related to increased thallium intake (Dolgner et al., 1983). Because thallium had been found to be developmentally toxic in laboratory animals (Karnofsky et al., 1950; Hall, 1972, 1976; Achenbach et al., 1979a,b; Gibson and Becker, 1970) (see also Section 10.4.2.1) and it was feared that it might also cause adverse developmental effects in humans, special care was applied to the investigation of the frequency of congenital malformations. Studies were performed in 300 children born in the area between January, 1978, and August, 1979. The mothers were asked to specify in detail occupational position, socioeconomic status, smoking and dietary habits, course of pregnancy and labor, as well as the child's birth weight and health status at birth. Urine and hair samples from the mothers were also collected and analyzed in order to get an estimate of the range of their past exposure. As a consequence of this investigation 19 children with suspected of congenital malformations and abnormalities were selected. After examination by a pediatrician, only 11 children were diagnoses as malformed or with abnormalities. In 2 of these children hereditary predisposition might have played an important role.

The alterations observed were classified as minor abnormalities (umbilical and inguinal hernia, hemangioma, dislocation of the hip) in six cases and major malformations (cleft lip, palate and jaw, microcephaly, lumbar meningomyelocele) in five cases. Because of the severity of the effects involved, the question of whether such congenital malformations could be attributed to effects of thallium was discussed by Dolgner et al. (1983). These authors pointed out several issues, namely:

1. The results seemed to suggest that in the cement plant area there was a significantly greater number of malformations than expected. However, in the Federal Republic of Germany there was no obligatory reporting of congenital malformations on birth certificates, and that might be the cause of incomplete data on the real frequency of congenital malformations, which could have been underestimated.
2. No specific pattern of congenital malformations was found.
3. There are several reports of human thallium intoxication during pregnancy with no consequences for the newborn (see Section 10.4.2.2)

In view of these considerations, Dolgner et al. (1983, p. 82) concluded that "a causal relationship between thallium exposure and congenital malformation among the children investigated in this study is unlikely to exist."

10.5. GONADOTOXICITY

10.5.1. Male Reproductive System

10.5.1.1. In Vitro

Primary cultures of Sertoli cells and germ cells treated with thallium concentrations corresponding to 35.7, and 1.4 mg Tl/testis, and observed 24, 48, and 72 h after treatment, showed a significant release of germ cells into the culture medium that was both concentration- and time-dependent (Gregotti et al., 1992). Changes in cell viability, protein content, glutathione (GSH) levels

(Fig. 1), and in the activity of selected testicular enzymes such as aldoso reductase, lactate deydrogenase, and pyruvate were also observed (Gregotti et al., 1991; 1993). These endpoints have been proposed as sensitive indices of altered Sertoli cells in vitro function (Williams and Foster, 1988).

Morphological investigation of cell cultures showed loss of germ cells and a significant decrease of the content of prepachytene and pachytene spermatocytes with alterations in Sertoli cell shape (Gregotti et al., 1992).

Inhibition of sperm motility was observed in in vitro experiments using boar ejaculates incubated with thallium concentrations lower than 0.001 mM (Manzo and Sabbioni, 1988).

10.5.1.2 In Vivo

The metabolic pattern of different thallium species was examined in the rat following the intraperitoneal administration of doses ranging from acute toxic doses to minute amounts, and the testis as well as the ovaries (Truhaut et al., 1957) were shown to be preferential sites for thallium accumulation, as listed in Table 2.

The metabolic pattern of thallium in the rat was independent of the dose absorbed, and the mechanism determining the disposition and tissue concentration of thallium operated at toxic and nontoxic doses (Sabbioni et al., 1980). Accumulation of thallium in the testis associated with morphological,

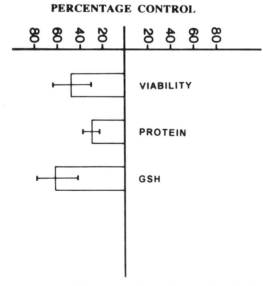

PERCENTAGE CONTROL

Figure 1. Effect on Sertoli cells of Tl treatment (0.5 μM) after 24 h. Viability assessed by neutral red dye assay. Protein values expressed as μg protein/dish. Total GSH and GSSG content assayed by enzyme recycling and expressed as nmol GSH/mg protein. Values are mean ± S.D. of determinations from three separate experiments; $p \leq 0.05$ (Gregotti et al., 1992; with the exception of GSH data that refer to unpublished findings).

Table 2   Tissue Distribution of ^{201}Tl in Rats Treated with 2 μg Tl/Rat[a]

	1.5 day	8 days
Kidney	4.43 ± 0.44	2.50 ± 0.68
Brain	0.16 ± 0.02	0.15 ± 0.03
Testes	0.69 ± 0.12	0.68 ± 0.22
Heart	0.37 ± 0.03	0.96 ± 0.06

[a] Values are expressed as percent dose/gram.
Source: Sabbioni and Manzo, 1980.

biochemical, and functional abnormalities in the reproductive tract was observed also in rats treated with 10 ppm thallium in drinking water for 2 months (Formigli et al., 1986).

Thallium treatment caused a significant decrease in sperm motility, and in most treated animals a high number of immature cells in their epididymal spermatozoa were present. Histological examination of the testis showed an increase of the release of later spermatids into the tubular lumen. Subsequent electron microscopy analysis revealed diffuse cytoplasmic vacuolation and distension of the smooth endoplasmic reticulum of the basal and perinuclear cytoplasm in Sertoli cells. Moreover the activity of β-glycuronidase, an enzyme primarily located in Sertoli cells, and spermatogonia were markedly decreased, as shown in Table 3.

Table 3   Effect of Thallium Administration on Male Reproductive System Parameters in the Rat

	Control	Thallium[a]
Relative testicular weight[b]	0.36 ± 0.02	0.37 ± 0.03
Sperm motility (%)	66.5 ± 0.02	40.15 ± 8.11[c]
Testicular enzyme activities		
β-glycuronidase[d]	0.046 ± 0.006	0.021 ± 0.004[c]
Acid phosphatase[e]	0.180 ± 0.027	0.140 ± 0.032
Sorbitol dehydrogenase[f]	4.000 ± 0.003	4.200 ± 0.001

[a] 10 ppm thallium as thallium sulfate in the drinking water for 60 days.
[b] Testis/body weight ratio × 100.
[c] Significant difference $p \leq 0.01$, as compared with paired controls ($n = 10$).
[d] nmoles phenolphtalein/min/mg protein.
[e] nmoles of p-nitrophenol/min/mg protein.
[f] nmoles of NAD converted to NADH/min/mg protein.
Source: Formigli et al., 1986.

Thallium has been shown to accumulate preferentially in the human testis (at 152 μg/g wet tissue) (Davis et al., 1981) and rat epididymis (at 16 ppm w/w) (Lameijer and Van Zwieten, 1977).

10.5.2. Female Reproductive System

Sexual activity has been reported to be decreased and estrus delayed in animals suffering from chronic thallium intoxication. The hypothesis proposed to explain these effects was that ovarian hormone production in mice was inhibited by thallium. Other studies failed to confirm this finding, and ovarian morphology was found to be normal in thallium intoxication (Smith and Carson, 1977). Degenerative changes in the female mice genital tract were seen after chronic thallium exposure (see Section 10.3).

10.5.3   Reproductive System: Human Data

Bonsignori and Scelsi (1970) reported an involvement of the endocrine glands such as the thyroid (alterations in basal metabolism), pancreas (hypoglycemia), testis (temporary impotence), and ovary (amenorrhea).

10.6.   CONCLUSIONS

Although studies of thallium toxicity have focused primarily on effects on the nervous system, skin, and cardiovascular tract, adverse effects on the reproductive tract function are now drawing more and more attention. In the literature there are numerous references to the mutagenic, carcinogenic, and teratogenic effects of thallium as well as on the influence of thallium on sexual behavior and reproduction (Smith and Carson, 1977). Thallium, administered during embryo development onto the chorioallantoic membrane of chick eggs, was embryotoxic. On the other hand, the evidence for thallium teratogenicity in mammal species is conflicting.

It has been hypothesized that the failure of thallium salts to produce severe teratogenic effects in mammals might be due to the relatively small amounts of thallium that can cross the placenta compared to the high doses injected into the egg, which is a closed system. In fact at the end of the incubation period, the chicks and the unabsorbed yolk sac contained the entire amount of the administered thallium (Gibson and Becker, 1970). Other authors (Ziskoven et al., 1983) observed that the difference in the nature and extent of possible fetal effects of thallium does not depend only on placental transfer and consequently on fetal thallium concentrations, but on obvious differences in strain-related thallium ion sensitivity.

There are a number of case reports of acute ingestion in humans during pregnancy with no mention of congenital malformations, even though the placental transfer of thallium to the fetuses has been demonstrated. Moreover, in epidemiological studies carried out in a population exposed long-term to low levels of thallium, no general increase of birth defects and no specific pattern of congenital malformations were found.

Although these experimental and clinical data suggest that thallium is probably not teratogenic to humans, it must be pointed out that the most severe effects of thallium toxicity are exerted on the nervous system and that neurological impairment is detectable only a few years after birth. Indeed, Galvan-Arzate and Rios (1993) found an increase of thallium concentrations in brain regions of newborn rats compared to 20-day-old rats possibly as a consequence of the incomplete formation of the blood–brain barrier, and suggested a higher susceptibility to the toxic effects of thallium as a function of age. As a proof, children followed up for 6 months to 7 years after thallium intoxication revealed residual neurological impairment described as "mental retardation" and "psychosis" (Reed et al., 1963).

Since in both rats and humans the pattern of thallium accumulation in the testis and in the brain appears to be similar, many studies on the male reproductive tract function have been carried out. These studies showed that the high accumulation of thallium in the testis is associated with morphological, biochemical, and functional abnormalities that are detectable earlier than any other signs of thallium toxicosis such as hair loss and peripheral nervous system disorders. The mechanism by which thallium affects the male reproductive system is not clear, but the involvement of Sertoli cells in the process underlying thallium-induced testicular toxicity was supported by in vitro studies. It is well known that in the male reproductive system the testes perform an important function in the process regulating the proliferation of spermatozoa and the production of hormones. One of the most important roles in spermatogenesis is played by Sertoli cells, which are intimately involved in the selective release of spermatids, in the disposition of residual bodies after spermeation, and in the formation of the blood–testis barrier. The peculiar thallium toxicity on the male reproductive system was confirmed by several investigations on the reproductive sequelae associated with paternal exposure to thallium, which found deleterious effects on sexual behavior and activities (Smith and Carson, 1977).

In conclusion, in humans there is no clear evidence of teratogenicity or genotoxicity of thallium after acute high-dose or chronic low-level exposure. Accidental, voluntary, or environmental exposure of females during pregnancy does not result in malformation of the newborn, even when placental transfer of thallium is proven. On the other hand, thallium accumulation in the testis with subsequent toxic manifestations on the male reproductive system has been demonstrated. In this respect, however, studies in men are still scanty, while it is well known that paternal exposure to toxicants can cause a wide range of adverse effects on reproduction. Effects of parental exposure to toxic agents before conception include reduced sexual activities and sperm fertility as well as unsuccessful fertilization or implantation.

In the light of these observations, more attention should be given to the toxic effects of thallium on the male reproductive system, as it appears to be an expecially sensitive organ for this metal's effects.

REFERENCES

Achenbach, C., Hauswirth, O., Heindrichs, C., Ziskoven, R., Kohler, F., Smend, J., and Kowalewki, S. (1979a). Toxizitat und Teratogenitat von Thallium. *Deutesches Arztebel., S,* 3189–3192.

Achenbach, C., Ziskoven, R., Kohler, F., Bahr, U., and Schulten, H. R. (1979b). Quantitative Spuren-analyse von Thallium in Biologischem Material. *Agnew. Chem.,* **91,** 944–945.

Anschuetz, M., Fierken, R., and Neubert, D. (1981). Studies on embryotoxic effects of thallium using the whole-embryo culture technique. In *Culture Technique: Symposium Prenatal Development, 5th,* Neubert, Diether, Merker, and Hans-Joachim de Gruyter (eds.). pp. 57–66.

Bonsignore, D., and Scelsi, R. (1970). L'intossicazione da tallio. *Folia Med.,* **9,** 188–204.

Cavanagh, J. B., and Gregson, M. (1978). Some effects of thallium salt on the proliferation of hair follicle cells. *J. Pathol.,* **125,** 179–191.

Cavanagh, J. B., Fuller, N. H., Johnson, H. R. M., and Rudger, P. (1974). The effects of thallium salts with particular references to the nervous system changes. *Q. J. MED.,* **43,** 293–319.

Cavieres, J. D., and Ellory, J. C. (1974). Thallium and the sodium pump in human red cells. *J. Phsysiol.,* **243,** 243–266.

Champy, C., Hatem, S., and Tharanne, J. (1958). Pathology of the genital tract of female mice after controlled thallium intoxication. *Compt. Rend. Soc. Biol.,* **152,** 906–907.

Claussen, U., Roll, R., Dolgner, R., Matthiaschik, G., Majewski, F., Stoll, F., and Rohrborn, G. (1981). Zur Mutagenicitat und Teratogenicitat von Thallium-unter Besonderer Berucksichtigung der Situation in Langerich. *Rhein Arztebl.,* **16,** 469–475.

Davis, L. E., Standefer, J. C., Kornfeld, M., Abercrombie, D. M., and Butler, C. (1981). Acute thallium poisoning: Toxicological and morphological studies of the nervous system. *Ann. Neurol.,* **10,** 38–44.

Di Nucci, A., Ferrini, B., Gregotti, C., Richelmi, P., and Manzo, L. (1979). Placental transfer of thallium and its modification by prussian blue. *IRCS,* **7,** 272.

Dixon, R. L., and Hall, J. L. (1982). Reproductive toxicology. *Principles and Methods in Toxicology,* Hayes, A. W. (ed.). Raven, New York, pp. 107–140.

Dolgner, R., Brockhaus, A., Ewers, U., Wiegand, H., Majewski, F., and Soddemann, H. (1983). Repeated surveillance of exposure to thallium in a population living in the vicinity of a cement plant emitting dust containing thallium. *Arch. Occup. Environ. Health.,* **52,** 79–94.

Douglas, K. T., Bunni, M. A., and Baindur, S. R. (1990). Thallium in biochemistry. *Int. J. Biochem.,* **22**(5), 429–438.

Erbsloh, J. (1960). Thalliumvergiftungen in der zweiten Schwangerschaftshalfte. *Arch. Toxicol.,* **18,** 156–159.

Formigli, L., Scelsi, R., Poggi, P., Gregotti, C., Di Nucci, A., Sabbioni, E., Gottardi, L., and Manzo, L. (1986). Thallium-induced testicular toxicity in the rat. *Environ. Res.,* **40,** 531–539.

Galvan-Arzate, S., and Rios, C. (1993). Thallium distribution in organs and brain regions of developing rats. *Toxicol.,* **90,** 63–69.

Gibson, J. E., and Becker, B. A. (1970). Placental transfer, embryotoxicity, and teratogenicity of thallium sulphate in normal and potassium deficient rats. *Toxicol. Appl. Pharmacol.,* **16,** 120–132.

Gibson, J. E., Sigdestad, C. P., and Becker, B. A. (1967). Placenta transport and distribution of thallium-204 sulphate in new born rats and mice. *Toxicol. Appl. Pharmacol.*, **10**, 408.

Governa, M., Scelsi, R., and Nicolò, G. (1969). Study on the action of thallium on human lynphocytes cultured in vitro. *Lav. Med.*, **6**, 240–245.

Gregotti, C., Faustman, E., and Manzo, L. (1991). Thallium induced changes in rat Sertoli cell function in vitro. *Pharmacol. Toxicol.*, **69**(II), 28.

Gregotti, C., Di Nucci, A., Costa, L. G., Manzo, L., Scelsi, R., Berté, F., and Faustman, E. (1992). Effects of thallium on primary cultures of testicular cells. *J. Toxicol. Environ. Hlth.*, **36**, 59–69.

Gregotti, C., Ponce, R., and Faustman, E. (1993). Mechanism of toxicity of Tl, B, and Pb on testicular cells in vitro. 32th Annual Meeting of Society of Toxicology, New Orleans, LA. (USA), Abstract p. 93.

Gunn, S. A., and Gould, T. C. (1970). Cadmium and other mineral elements. In Johnson, A. D., Gomes, W. R., and Vandermark, L. H. (eds.). *The Testis*. Academic, New York, Vol. 3, pp. 378–482.

Hall, B. K. (1972). Thallium-induced achondroplasia in the embryonic chick. *Dev. Biol.*, **28**, 47–60.

Hall, B. K. (1976). Thallium-induced achondroplasia in chicken embryos and the concept of critical periods during development. *Teratology*, **15**, 1–16.

Hart, M. M., and Adamson, R. H. (1971). Antitumor activity and toxicity of salts of inorganic group IIIa metals: Aluminium, gallium, indium, and thallium. *Proc Nat. Acad. Sci. USA*, **68**, 1623–1626.

Johnson, W. (1960). A case of thallium poisoning during pregnancy. *Med. J. Austral.*, **II**, 540–542.

Karnofsky, D. A., Rigdway, L. P., and Patterson, P. A. (1950). Production of achondroplasia in the chick embryo with thallium. *Prof. Soc. Exp. Biol. Med.*, **73**, 255–259.

Kassis, A. I., Adelstein, S. J., Haydock, C., and Sastry, K. S. R. (1983). Thallium-201: An experimental and theoretical radiobiological approach to dosimetry. *J. Nucl. Med.*, **26**, 59–67.

Kelsey, K. T., Donohoe, K. J., Baxter, B., Memisoglu, A., Little, J. B., Caggana, M., and Liber, H. L. (1991). Genotoxic and mutagenic effects of the diagnostic use of thallium-201 in nuclear medicine. *Mut. Res.*, **260**, 239–246.

Kunze, M. (1972). Der Einflusse von Thalliumazetat auf das Wachstum von Acholeplasmateceae, Mycoplasmataceae und einigen Bakterienspezies. *Zbl. Bakt., I. Abt. Orig. A.*, **222**, 535–539.

Lameijer, W., and Van Zwieten, P. A. (1976). Acute cardiovascular toxicity of Thallium (I) ions. *Arch. Toxicol.*, **35**, 49–61.

Lameijer, W., and Van Zwieten, P. A. (1977). Kinetic behavior of thallium in the rat. *Arch. Toxicol.*, **37**, 265–273.

Landauer, W. (1960). Experiment concerning the teratogenic nature of thallium: Polyhydroxy compounds, histidine and imidazole as supplements. *J. Exp. Zool.*, **143**, 101–105.

Manzo L., and Sabbioni, E. (1988). Thallium. In *Handbook on Toxicity of Inorganic Compounds*. Seiler, H. G., Sigel, H., and Sigel, A. (eds.). Marcel Dekker, New York, pp. 677–688.

Manzo, L., Scelsi, R., Moglia, A., Poggi, P., Alfonsi, E., Pietra, R., Mousty., F., and Sabbioni, E. (1983). Long-term toxicity of thallium (I) in the rat. In *Chemical Toxicology and Clinical Chemistry of Metals*. Brown S. S, and Savory, J. (eds.). Academic, Orlando/London, pp. 401–405.

Moeschlin, S. (1980). Thallium poisoning. *Clin. Toxicol.*, **17**, 133–146.

Olsen, I., and Jonsen, J. (1982). Whole-body autoradiography of 204-Tl in embryos, fetuses and placenta of mice. *Toxicol.*, **23**, 353–358.

Petersohn, K. L. (1960). Thalliumvergiftungen in der Schwangerschaft. *Arch. Toxicol.*, **18**, 160–164.

Prick, J. J. G., Sillevis Smitt, W. G., and Muller, L. (1955). *Thallium Poisoning*. Elsevier, Amsterdam, pp. 1–55.

Prinz, B., Krause, G. H. M. Stratmann, H. (1979). Thalliumschaden in der Umgebung de Dyckerhoff Zementwerke AG. *Langerich, Westfalen, Staub Reinhalt Luft*, **39**, 457–462.

Rao, D. V., Govelitz, G. F., and Sastry, K. S. R. (1983). Radiotoxicity of thallium-201 in mouse testes; inadequacy of conventional dosimetry. *J. Nucl. Med.*, **24**, 145–153.

Reed, D., Crawley, J., Faro, S. N., Pieper, S. J., and Kurland, L. T. (1963). Thalliumtoxicosis. Acute manifestations and sequelae. *JAMA*, **183**, 516–522.

Sabbioni, E., and Manzo, L. (1980). Metabolism and toxicity of thallium. In *Advances in Neurotoxicology*, Manzo, L., Lacasse, Y., Lery, M., Roche, L. (eds.). Pergamon, Oxford/New York, pp. 249–270.

Sabbioni E., Marafante, E., Rade, J., Di Nucci, A., Gregotti, C., and Manzo, L. (1980). Metabolic patterns of low and toxic doses of thallium in the rat. In *Mechanism of Toxicity and Hazard Evaluation*, Holmsted, B., Lauwerys, R., Mercier, M., Robertfroid, M. (eds.), Elsevier, Holland, pp. 559–564.

Smith, I. C. H., and Carson, B. L. (1977). *Trace Metals in the Environment. Thallium*, Vol. 1, Ann Arbor Science, Ann Arbor, MI.

Tandon, S. P., and Mishra, M. M. (1968). Effect of some rare elements on nitrification by nitrate-forming bacteria in soil suspension. *Zentralb. Bakteriol. Parsitenk. Infektionskr. Hyg. Abt. II*, **122**, 155–157.

Truhaut, R., Banquet, P., and Capot, L. (1957). Sur la repartition du thallium radioactif chez le rat. *Compt. Ren.*, **245**, 116–119.

Williams, J., and Foster, P. M. (1988). The production of lactate and pyruvate as sensitive indices of altered Sertoli cells in vitro following the addition of various testicular toxicants. *Toxicol. Appl. Pharmacol.*, **94**, 160–170.

Zashukina, G. D., Krasovsky, G. N., Vasilyeva, I. M., Sdirkova N. I., Solokovsky, V. V., Kenesariev, U. I., and Vasyukovich, L. Y. (1980). Approach to the determination of a mutagenic potential of environmental pollutants as illustrated by the detection of the mutagenic effect of thallium carbonate. *Byull. Eksp. Biol. Med.*, **90**, 723–726.

Zashukina, G. D., Vasilyeva, I. M., Sdirkova N. I., Krasovsky, G. N., Vasyukovich, L. Y., Kenesariev, U. I., and Butenko, P. G. (1983). Mutagenic effect of thallium and mercury salts on rodent cells with different repair activities. *Mutat. Res.*, **124**, 163–173.

Ziskoven, R., Achenbach, C., Schulten, H. R., and Roll, R. (1983). Thallium determination in fetal tissues and maternal brain and kidney. *Toxicol. Lett.*, **19**, 225–231.

Zitko, V. (1975). Toxicity and potential pollution of thallium. *Sci. Total. Environ.*, **12**, 157–160.

11

AGE-DEPENDENT DIFFERENCES IN THE NEPHROTOXICITY OF THALLIUM(I) SULFATE IN RATS

Dorothea Appenroth

Institute of Pharmacology and Toxicology, Friedrich Schiller University, D-07740 Jena, Germany

Stephan Gambaryan

Institute of Evolutionary Physiology and Biochemistry of the Russian Academy of Science, St. Petersburg, Russia

11.1. **Introduction**
    11.1.1. Use and Biological Effects of Tl Compounds
    11.1.2. Biochemical Basis of Tl Toxicity
    11.1.3. Developmental Aspects of Nephrotoxicity
11.2. **Kinetics of Tl Assimilation**
11.3. **Effects of Tl on the Kidney**
    11.3.1. Effects on Renal Morphology
            11.3.1.1. Diuretic Effect of Furosemide
    11.3.2. Effects on Renal Excretory Functions

Thallium in the Environment, Edited by Jerome O. Nriagu.
ISBN 0-471-17755-5. © 1998 John Wiley & Sons, Inc.

11.1.   INTRODUCTION

11.1.1.   Use and Biological Effects of Tl Compounds

Thallium was discovered in 1861 by William Crookes. It is not a normal constituent of tissues and has no physiological role. Thallium was used in the late 1800s for the treatment of syphilis, gonorrhea, gout, dysentery, and night sweats from tuberculosis (Saddique and Peterson, 1983). Dermatologists used salts of thallium also as a depilatory agent. Numerous unpleasant effects precluded this wide acceptance in medicine. Currently, the only medical application is the nonpharmacological use of ^{201}Tl as a tracer in myocardial imaging (Kelner, 1994). Use of Tl as pesticide started in 1920. Due to its contribution to severe intoxication, this application of thallium was rejected by the WHO (1973). Thallium is used industrially in small quantities in optical glasses, photographic films and their development, and in special power systems (Chandler and Scott, 1986). Lately thallium has been used as a tool in several aspects of biochemistry (Douglas et al., 1990).

Thallium occurs as Tl(I) and Tl(III) of which in aqueous solution the Tl(I) is the more stable and also more toxic (Douglas et al., 1990; WHO, 1996). Thallium (I) possesses an ionic radius similar to potassium, which is one important reason for its toxicity. No data about the biotransformation of thallium are available (WHO, 1996). In mammalian species, the acute LD$_{50}$ for Tl compounds for all routes of administration ranges between about 5 and 70 mg/kg body weight (b.wt., Kazantzis, 1986). Symptoms of Tl intoxication in humans develop within a few days after Tl administration, and include gastrointestinal manifestation, nervous system dysfunction, affection of skin, and cardiovascular system (Sabbioni and Manzo, 1979; Saddique and Peterson, 1983; Chandler and Scott, 1986). Many reports exist that deal with these toxic effects (for a review see Moeschlin, 1980; Saddique and Peterson, 1983; Chandler and Scott, 1986). But nephrotoxic effects are only mentioned casually, such as oliguria and albuminuria in human intoxications (Shabalina and Spiridonowa, 1979; Moeschlin, 1980; Saddique and Peterson, 1983; Chandler and Scott, 1986; Aoyama et al., 1986). Few studies exist that deal with morphological changes in kidney following Tl poisoning in experimental animals (Herman and Bensch, 1967; Danilewicz et al., 1979; Appenroth et al., 1995, 1996). Nephrotoxic effects of acute Tl administration were investigated more in detail in adult and developing rats by Appenroth et al. (1995, 1996).

11.1.2. Biochemical Basis of Tl Toxicity

Thallium is recognized to be one of the most toxic heavy metals. Its ability to interfere with a variety of potassium-dependent processes is thought to play a significant role (Mulkey and Oehme, 1993; Kelner, 1994). Like K it depolarizes membranes and reverses cardiac effects of hypokalemia (Saddique and Peterson, 1983). Its affinity to Na^+/K^+-ATPase is 10 times greater than that of potassium (Douglas et al., 1990). Thallium is actively accumulated into mitochondria and appears to act as an uncoupler of oxydative phosphorylation (Melnick et al., 1976). Thyresson (1950) showed by in vitro investigations that aerobic respiration of skin, brain, and kidney tissue was markedly inhibited by Tl and that the inhibition of anaerobic glycolysis was approximately one third that of aerobic respiration. It has been suggested that Tl impairs cellular energy metabolism and inhibits the activity of FAD enzymes such as GSSG reductase by causing a deficiency of riboflavin as substrate or cofactor, respectively (Cavanagh, 1991; Mulkey and Oehme, 1993). Furthermore, thallium has a high affinity to SH groups (Bugarin et al., 1989; Douglas et al., 1990), which are structurally important in several classes of enzymes. This holds true only for Tl(III) compounds (Douglas et al., 1990). However, the precise biochemical mechanisms underlying the clinical manifestations of Tl toxicity are yet to be proven. In a recent paper (Appenroth and Winnefeld, 1998) we reconsidered some hypotheses on the mechanism of toxic thallium action discussed in the literature.

11.1.3. Developmental Aspects of Nephrotoxicity

The kidney is the main excretory organ, and therefore many final products of metabolism and most of foreign substances such as drugs and environmental chemicals (e.g., Tl compounds) pass through the kidney. Although the net result consists of their excretion, physiological processes may lead to temporary accumulation of xenobiotics in renal tissue thereby causing toxic effects. It is known that the renal function of rats is still immature after birth (Kersten and Bräunlich, 1968; Appenroth and Bräunlich, 1986). This is true for the glomerular filtration rate (GFR; Appenroth et al., 1996a) as well as for active tubular functions such as electrolyte transport and the secretion of organic acids and bases. Moreover, intrarenal biotransformation may be different in young and adult rats. With respect to the developmental stage of these functions, 10-day-old rats are young and immature, whereas 55-day-old rats are mature adults. As already shown for cisplatin (Appenroth and Bräunlich, 1984; Appenroth et al., 1990), chromate (Appenroth and Bräunlich, 1988), and Tl (Appenroth et al., 1996a), immature kidney function is of importance for the degree of nephrotoxicity.

11.2. KINETICS OF Tl ASSIMILATION

Rapid and complete absorption, distribution, and retention in all tissues and a rather slow elimination are the basic aspects of Tl metabolism. Lie et al. (1960) administered ^{204}Tl by six different routes. Whatever the route of administration, the body burden was similar, that means Tl was always completely absorbed (Lie et al., 1960). After intraperitoneal (i.p.) injection we determined maximal Tl concentrations in serum 1 h after the administration (Fig. 1). Absorption seems to be retarded in 10-day-old compared with adult rats. We determined Tl concentration in serum excluding the part that was located in red blood cells. Because no binding of Tl to plasma proteins could be detected (Kelner, 1994) the Tl, determined in serum, represents the free Tl portion that can be distributed in many organs. The relative distribution volume, given in the literature as 5–6 L/kg (Saddique and Peterson, 1983), was calculated in our rats to be equal in both age groups (10-day-old rats 2.57 L/kg, 55-day-old rats 2.70 L/kg). In the literature serum half-life time ($t/2$) in whole blood varied from 1.5 to 4 days (Kelner, 1994), which corresponds with 32.2 h determined in our adult rats (Fig.1). It is approximately three times longer in young rats (103.4 h). This is shown to be caused by lower excretion via urine in 10-day-old rats (Fig. 1). We did not determine the portion of Tl excreted via feces (approximately double that excreted via urine; Kelner, 1994; Lehmann and Favari, 1985), which might contribute to differences in Tl half-life time in serum. However, prolonged half-life time in young rats can be explained satisfactorily by minor Tl excretion via urine.

Since Tl and K are monovalent ions with similar ionic radii, Tl interferes with K-dependent processes and mimics K in its movement and intracellular accumulation in mammals (Gehring and Hammond, 1967). This holds true also for the passage of Tl through the kidney. Like K it is glomerularly filtered, tubularly secreted, and 50% of Tl is reabsorbed (Lund, 1956). After the administration of Tl, young rats excreted significantly less Tl than adult rats did (Fig. 1). This was in parallel to the age differences of K excretion described in the literature (Kersten and Bräunlich, 1968) and to the K excretion of young and adult control rats shown in Figure 8.

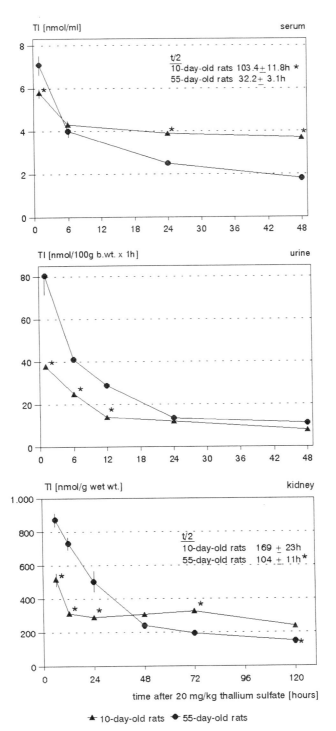

Figure 1. Kinetics of Tl in serum, urinary excretion, and time course of Tl concentration in renal tissue of 10- and 55-day-old rats after one single administration of 20 mg/kg b.wt. thallium sulfate intraperitoneally (i.p.). *, statistically significant differences between 10- and 55-day-old rats ($n = 6$, $p \leq 0.05$).

As described in the literature (Lund, 1956; Sabbioni and Manzo, 1979; Leloux et al., 1987), after acute intoxication highest organ Tl concentrations were detected in renal tissue. As shown in Figure 1 already 1 h after administration, maximal Tl concentrations were measured in kidney tissue (without separation between cortex and medulla). In comparison to serum, Tl was enriched in the kidney approximately 100-fold. During the first 24 h after administration, Tl concentrations are significantly lower in 10- than in 55-day-old rats. Thallium excretion as well as its enrichment in several organs such as renal tissue is connected with the activity of the Na^+/K^+-ATPase, which is significantly lower in whole renal tissue of 10- than in 55-day-old control rats (Bräunlich et al., 1989), causing lower Tl concentrations in these young animals. The $t/2$ of Tl is more than 50% longer in young rats (Fig. 1), although the volume of distribution was calculated to be equal in young (2.57 L/kg) and adult rats (2.70 L/kg). The following explanations exist: (1) less urinary excretion of Tl (Fig. 1); (2) Tl itself inhibits the activity of the ATPase in 10-day-old rats (Table 1, second day after Tl administration), thereby decreasing its own excretion; (3) it may be possible that the binding of Tl to renal tissue is stronger in young rats (no data available).

The difference in concentrations of Tl between cortex and medulla (Table 1) can be explained by the different activities of Na^+/K^+-ATPase in these regions of the kidney. The data of Careaga-Olivares and Morales-Aguilera (1990) obtained in rabbits correspond with our data. The consequences of different Tl concentrations in renal medulla and cortex on the extent and localization of nephrotoxic effects will be described below.

11.3.   EFFECTS OF Tl ON THE KIDNEY

11.3.1.   Effects on Renal Morphology

Histological necropsy findings in fatal human Tl ingestion showed tubular necrosis (Hologgitas et al., 1980), extensive in kidney cortex (Cavanagh et al., 1974). Herman and Bensch (1967) and Danilewicz et al. (1979) reported about morphologic destructions in the proximal and distal tubules of acutely Tl poisoned rats. Our own morphologic investigations were done in 10- and 55-day-old rats at different times after one single dose of 20 mg Tl_2SO_4/kg b.wt. In adult rats 12 h after Tl administration destructions in the renal medulla occurred mainly in the thick ascending limb (TAL) of the loop of Henle (Fig. 2A and 2B). The maximum of destructions were demonstrated at the second day after Tl administration (Fig. 2C and 2D). Cells of TAL appeared to be swollen, and their cytoplasma contained electron transparent vacuoles mostly in the central part of the cells. The lumina of some tubules were filled with picnotic nuclei and cell debris. The structure of proximal tubules and glomeruli was not affected morphologically (Appenroth et al., 1995), which was confirmed by testing the activity of the carrier system for organic acids, located

Table 1 Concentration of Tl (nmol/g wet wt.) and Activity of Na$^+$/K$^+$-ATPase (nmol P/mg prot. × min) in Renal Cortex and Medulla of 10- and 55-Day-Old Rats at the Second and Fifth Day after One Single Administration of 20 mg/kg b.wt. Thallium Sulfate (i.p.) in Comparison to Control Rats

	Control		Second Day after Tl		Fifth Day after Tl	
	Cortex	Medulla	Cortex	Medulla	Cortex	Medulla
Tl						
10-day-old	0	0	440 ± 21	713 ± 34[a]	332 ± 19	610 ± 35[a]
55-day-old	0	0	429 ± 89	1525 ± 94[a,b]	163 ± 9[b]	281 ± 31[a,b]
ATPase						
10-day-old	34 ± 2	34 ± 2	33 ± 2	22 ± 1[a,c]	34 ± 1	42 ± 2[a,b,c]
55-day-old	60 ± 3[b]	112 ± 4[a,b]	66 ± 3[b]	144 ± 4[a,b,c]	54 ± 4[b]	98 ± 6[a,b]

[a] Statistically significant differences between cortex and medulla.
[b] Statistically significant differences between 10- and 55-day-old rats.
[c] Statistically significant differences in comparison to control ($n = 6$, $p \leq 0.05$).

221

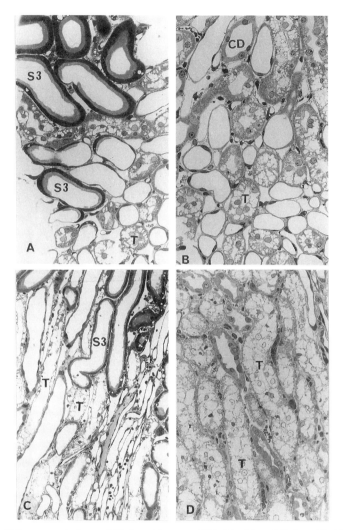

Figure 2. Kidney morphology of 55-day-old rats at 12 h (A, B) and 2 days (C, D) after the administration of one single dose of 20 mg/kg b.wt. thallium sulfate. T, thick ascending limb of the loop of Henle (TAL); CD, collecting duct; S3–S3, segment of proximal tubule. Magnification: A, B × 360, C × 400.

in proximal tubules in renal cortical slices, in vitro. The accumulation of *p*-aminohippurate (model substances for the transport function of this carrier) was not impaired after Tl administration (data not shown). This was in contradiction to Herman and Bensch (1967) who found destructions in proximal tubules and glomeruli. In our opinion this discrepancy is caused by the methods of fixation used. These authors needed 2–4 min after killing the animals for placing the tissue into the fixative. We used the method of perfusion fixing in situ, which prevented hypoxic damage of highly susceptible proximal tubules (Appenroth et al., 1995). Concerning the destruction of TAL, our findings are in agreement with the results of Danilewicz et al. (1979). At the fifth day after Tl administration, glomeruli and proximal tubules were comparable to controls (Appenroth et al., 1995). At that time regenerating processes started in the medulla. In the inner and outer stripes of the outer medulla, some of the TAL cells consisted of a thin layer of epithelium with prominent nuclei directed into the lumen, which were filled by products of cell necrosis (Appenroth et al., 1995). Some other TALs consisted of cells in different stages of regeneration. The cytoplasma of these cells possessed large electron transparent vacuoles. At the tenth day after Tl administration, the structures of kidney cortex and medulla were absolutely comparable to controls (Appenroth et al., 1995). The four times higher concentration of Tl in renal medulla in comparison to cortex (Table 1) underlines our morphologic findings. The results of morphological investigations in 10-day-old rats are shown in Figure 3. Experiments were also done at different times after one single dose of 20 mg Tl$_2$SO$_4$/kg b.wt. As shown in Figure 3, there were absolutely no structural changes at the fifth day after Tl administration. As already described for adult rats, in 10-day-old rats the accumulation of p-aminohippurate was not influenced by Tl (data not shown) supporting normal structure of proximal tubules. Despite missing structural changes in the nephron, at this time

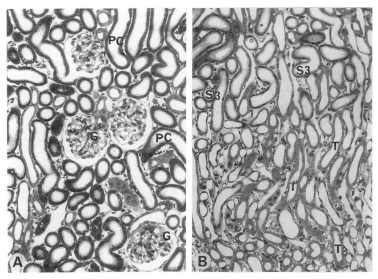

Figure 3. Kidney morphology of 10-day-old rats at the fifth day after one single dose of 20 mg Tl$_2$SO$_4$/kg b. wt. G, glomerulum; PC, proximal convoluted tubule; S3–S3, segment of proximal tubule; T, thick ascending limb of the loop of Henle (TAL). Magnification $\times$ 260.

significant disturbances appeared in renal excretory function as will be described below. Investigations on days 2 and 10 after Tl administration also did not exert any morphological changes in 10-day-old rats (Appenroth et al., 1996a).

In summary, there were severe destructions of TALs in adult rats at the second day after Tl administration, which regenerated completely up to the tenth day, whereas in young rats renal structure remained completely unaffected by Tl administration. The lower concentration of Tl in 10-day-old rats during the first 24 h in whole renal tissue (Fig. 1), and in the medulla at least during 2 days after Tl administration (Table 1) seems to be a determinant for missing morphological changes in these young rats. Celsi et al. (1986) have shown that after nephrectomy compensatory renal growth was more active in the remaining kidney of young rats, which was attributed to the higher mitotic capacity of these young compared with adult rats. Therefore it might be possible that slight effects appeared but were repaired immediately.

11.3.1.1. Diuretic Effect of Furosemide

Furosemide acts on TAL, which we found to be morphologically destroyed by Tl in adult rats only. Therefore we used this diuretic as a tool to search for functional disturbances because of the missing morphologic effects of Tl in young rats (Fig. 3; Appenroth et al., 1996b). The diuretic effect of furosemide in 10- and 55-day-old control and Tl-treated rats is shown in Figure 4. When Tl was administered to 10-day-old rats, the diuretic effect was significantly diminished from day 5 up to day 10 after Tl, as it was in 55-day-old rats on day 2. From these results we conclude that, although no morphologic changes were demonstrable in the TALs of young rats, there occurred functional changes in the same region of kidney as in adult rats.

11.3.2. Effects on Renal Excretory Functions

As a general symptom of acute Tl toxicity, Herman and Bensch (1967) mentioned marked loss of body weight after the administration of 20–50 mg thallium acetate/kg b.wt. This finding could not be confirmed by our results, neither in young nor in adult rats (Fig. 5).

With the exception of our previous studies (Appenroth et al., 1995, 1996a), systematic investigations of Tl effects on renal excretory function are not available. Mulkey and Oehme (1993) reviewed that renal function is usually not grossly impaired, despite the fact that the kidneys accumulate the highest concentration of Tl of any organ. On the other hand, descriptions of human Tl intoxications report changes in renal function, such as oliguria, diminished creatinine clearance, raised blood urea nitrogen (BUN), and albuminuria

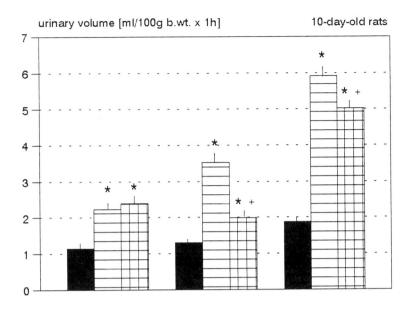

urinary volume [ml/100g b.wt. x 1h]   10-day-old rats

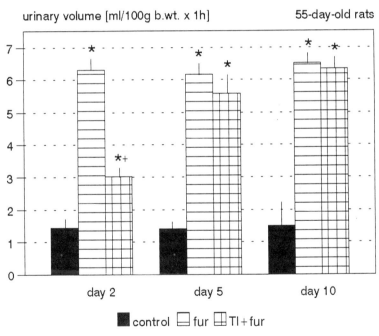

urinary volume [ml/100g b.wt. x 1h]   55-day-old rats

day 2   day 5   day 10

■ control ⊟ fur ⊞ Tl + fur

Figure 4. Diuretic effect of furosemide (30 mg/kg b.wt., i.p.) in 10- and 55-day-old control and Tl-treated rats at different times after one single administration of 20 mg/kg b.wt. thallium sulfate i.p. *, statistically significant differences between control- and furosemide- (fur) and Tl + furosemide- (Tl + fur) treated rats ($n = 6, p \leq 0.05$); +, statistically significant differences between furosemide and Tl + furosemide-treated rats ($n = 6, p \leq 0.05$).

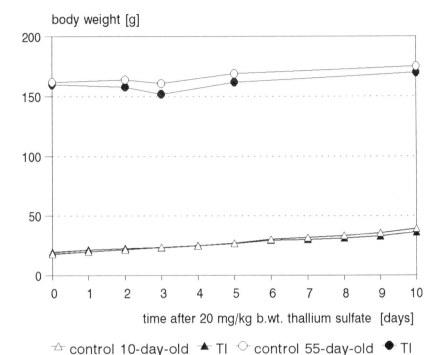

Figure 5. Body weight gain in 10- and 55-day-old rats during 10 days after one single administration of Tl.

(Shabalina and Spiridonowa, 1979; Sabbioni and Manzo, 1979; Moeschlin, 1980; Saddique and Peterson, 1983; Chandler and Scott, 1986; Aoyama et al., 1986). Our results (Figs. 6–8) defined more clearly these symptoms concerning the time course and the influence of age. In 55-day-old rats Tl produced a strong oliguric effect during the first 3 days after one single dose of Tl (Fig. 6). At this time, tubular lumina were filled with pycnotic nuclei and cell debris (Fig. 2), hindering the urine flow. Additionally, GFR was significantly decreased, which was reflected by a significantly increased concentration of BUN showing a corresponding time course to GFR (Fig. 7). This cannot be explained by glomerular destruction (Appenroth et al., 1995). Direct effects of Tl on blood vessels were not known. In human beings tachycardia and increased blood pressure occurred in the second week after Tl intoxication (Moeschlin, 1980). It may be due to the stimulating effect of Tl on ATPase in the chromaffin cells leading to increased output of catecholamines (Moeschlin, 1980). The same mechanism may also cause transient vasoconstriction in the vasa afferentia, thereby significantly decreasing GFR. As known from rats (Kersten and Bräunlich, 1968; Appenroth and Bräunlich, 1988), urine excre-

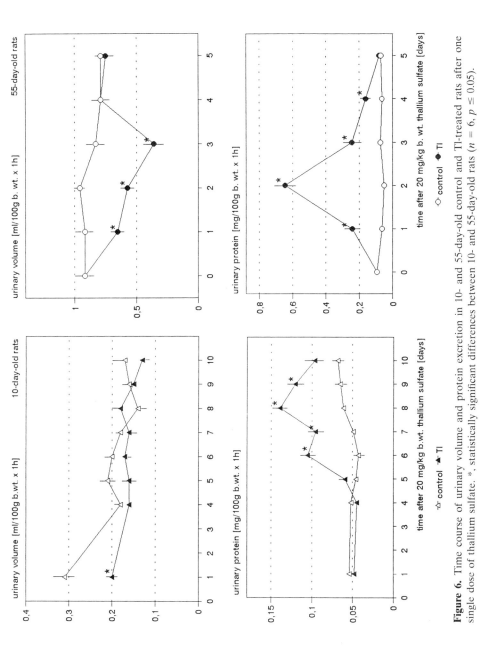

Figure 6. Time course of urinary volume and protein excretion in 10- and 55-day-old control and Tl-treated rats after one single dose of thallium sulfate. *, statistically significant differences between 10- and 55-day-old rats ($n = 6$, $p \leq 0.05$).

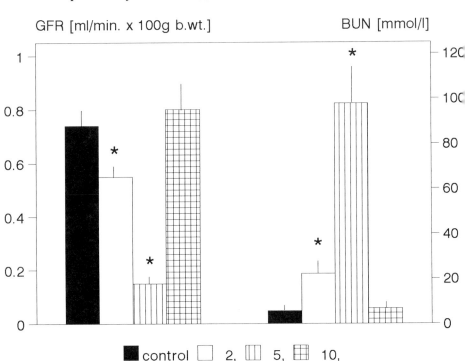

■ control ☐ 2, ⬚ 5, ⊞ 10,

time after 20mg/kg b.wt.thallium sulfate [days]

Figure 7. Glomerular filtration rate (GFR) and concentration of blood urea nitrogen (BUN) in 55-day-old control and Tl-treated rats at different times after the administration of one single dose of thallium sulfate. *, statistically significant differences between control- and Tl-treated rats ($n = 6$, $p \leq 0.05$).

tion was significantly lower in young control rats. Oliguria in 10-day-old rats was of a less degree compared with 55-day-old rats. This is consistent with lacking morphologic changes (Fig. 3) and normal GFR (Appenroth et al., 1996a) in 10-day-old rats at any time after the administration of Tl.

Proteinuria is a sensitive parameter of nephrotoxicity for many types of nephrotoxins, such as drugs and chemicals (Bernard et al., 1991), and it is suitable for nephrotoxicity studies in young rats, too (Appenroth and Bräunlich, 1984, 1988). Proteins excreted via urine are glomerularly filtered mainly in dependence on molecular weight, and more than 90% are reabsorbed in proximal tubules. We determined urinary protein excretion on the whole without differentiation between certain protein species. In 55-day-old rats maximal proteinuria was sixfold in comparison to controls at the second, and normalized at the fifth, day after Tl administration. Age differences in proteinuria of control rats were described previously (Appenroth and Bräun-

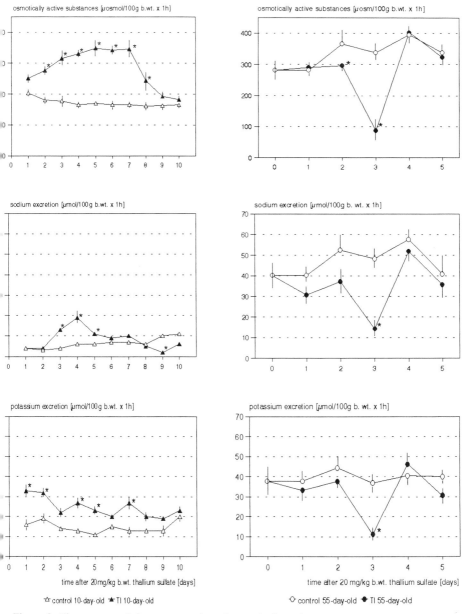

Figure 8. Time course of the concentration of osmotically active substances as well as of the urinary excretion of sodium and potassium in 10- and 55-day-old control- and Tl-treated rats after one single dose of thallium sulfate. *, statistically significant differences between control and Tl-treated rats ($n = 6$, $p \leq 0.05$).

lich, 1988) and confirmed by the results presented in Figure 6. Proteinuria after Tl administration to 10-day-old rats was enhanced only two- to threefold (Fig. 6), which corresponds well with missing morphologic changes (Fig. 3) and lower Tl concentration in renal tissue of these rats (Fig. 1, Table 1). However, significant proteinuria in 10-day-old rats occurred some days later than in 55-day-old rats. The reason for this shift in time course is not known. Enhanced proteinuria indicates disturbances in glomerular filtration and/or the reabsorption of proteins in the proximal tubule. As we have shown (Figs. 2 and 3), there was no destruction neither in glomeruli nor in proximal tubules of 10- and 55-day-old rats. Nevertheless, enhanced proteinuria proved functional impairment of glomeruli and/or proximal tubuli in young as well as adult rats.

The passage of Tl through the kidney is quite similar to that of K. Potassium chloride administered in pharmacological doses has been used successfully to increase the excretion of Tl by its competition with Tl for reabsorption sites in the distal tubule, including TAL (Gehring and Hammond, 1967; Saddique and Peterson, 1983). On the other hand, no data are available concerning the influence of Tl on the excretion of K or renal electrolyte excretion in general. In our experiments the excretion of the total amount of osmotically active substances in 55-day-old rats was decreased significantly on day 3 after Tl administration (Fig. 8). This is approximately the sum of diminished excretion of Na and K and associated anions at this time. Potassium is cotransported with Na from the lumen into the cell of TALs. The main portion of reabsorbed K returns back, to the tubular fluid by diffusion. TAL was shown (Fig. 2) to be destroyed strongly at the second day after Tl in 55-day-old rats, and signs of restitution were shown (Fig. 2). Perhaps, the flow of K back to tubular fluid is predominantly disturbed, which might be the reason for decreased K excretion on day 3 after Tl administration (Fig. 8). As shown in Table 1 despite severe morphologic destruction, the activity of Na^+/K^+-ATPase was increased significantly in renal medulla at the second day after Tl enhancing the transport of Na out of the cells into the interstitium, which could explain decreased urinary Na excretion at the second day after Tl administration (Fig. 8).

The age differences of electrolyte excretion in control rats are in agreement with literature data (Kersten and Bräunlich, 1968; Aperia and Elinder, 1981). The influence of Tl on the excretion of osmotically active substances Na and K in 10-day-old rats differs completely from that of adult rats (Fig. 8). The excretion of osmotically active substances is increased significantly from the second up to eight day after Tl administration. In contrast to 55-day-old rats both sodium and potassium excretions were enhanced. Although there were no structural changes in TALs in 10-day-old rats (Fig. 3), the diminished effectiveness of furosemide demonstrated functional changes in this part of the nephron. Therefore it is probable that the impairment of Na/K cotransport at the luminal side of the TAL is the reason for the enhancement of Na and K excretion in 10-day-old rats as the consequence of Tl administration.

11.3.3. Effects on Na⁺/K⁺-ATPase in Renal Tissue

Thallium had been shown to replace K in the activation of Na^+/K^+-ATPase and was unique in possessing an affinity for the K-activating site of this enzyme 10 times greater than K (Gehring and Hammond, 1967; Britten and Blank, 1968). However, this activity was inhibited completely by high levels of Tl (Mulkey and Oehme, 1993). Na^+/K^+-ATPase is responsible for the active cation transport in cell membranes and plays an important role in renal electrolyte excretion. Its activity in whole kidney was shown to develop postnatally (Table 1; cf. Bräunlich et al., 1989). Moreover, in 10-day-old rats the activities still were equal in cortex and medulla, whereas the medulla of 55-day-old rats possessed approximately double the activity of the cortex. Although at the second day after the administration of Tl there are extensive destructions in the renal medulla of 55-day-old rats (Fig. 2), a significant stimulation of the activity of Na^+/K^+-ATPase was detectable (Table 1), which returned to control value at the fifth day. Since Tl is able to compete for K at its binding site, the increased activity of Na^+/K^+-ATPase probably contributes to the accumulation of Tl in renal medulla thereby promoting Tl nephrotoxicity. At the same time ATPase activity is significantly decreased in the medulla of 10-day-old rats, which may contribute to lower the Tl concentration in the medulla compared with adult rats. At the fifth day ATPase activity was significantly enhanced in the medulla of young rats. This can explain the retarded decline of the Tl concentration in renal tissue of 10- compared with 55-day-old rats (Fig. 1). Concerning the different effect of Tl on the activity of Na^+/K^+-ATPase in relation to time after Tl administration, it can be speculated that this enzyme possesses different properties in 10- and 55-day-old rats.

The influence of Tl on the activity of Na^+/K^+-ATPase was discussed to be one important part of the molecular basis of its toxicity (Mulkey and Oehme, 1993). This cannot be supported because our results did not show any correlation to the changes of renal electrolyte excretion as a consequence of Tl, administration in 10- as well as in 55-day-old rats. We conclude that the activity of Na^+/K^+-ATPase is an indirect factor in Tl nephrotoxicity by influencing Tl concentration in renal tissue.

11.3.4. Effects on GSSG Reductase Activity in Renal Tissue

GSSG reductase (GR) is a dimeric FAD-containing enzyme that assists in the maintenance of the intracellular levels of glutathione (Bates, 1993; Gilliland, 1993). It catalyzes the reaction that converts oxidized glutathione into the reduced form (GSH). GSH acts as a reducing agent and antioxidant, participates in detoxification reactions of many substances, and is implicated in many other cellular functions (Deneke and Fanburg, 1989). One basic mechanism of Tl toxicity discussed is its ability to cause a deficiency of riboflavin and riboflavin-derived cofactors (e.g., in GR) acting as flavoprotein inhibitor (Mulkey and Oehme, 1993). The GSH redox system is highly active in renal tissue (Dillard et al., 1987), and therefore we determined the activity of GR in renal tissue of 10- and 55-day-old rats at different times after Tl administration (Fig. 9). In Tl-treated rats of both age groups the activity of GR was elevated significantly, more expressed in 10- than 55-day-old rats. These results show that in renal tissue Tl does not act as a flavoprotein inhibitor, at least in case of GR. Aoyama et al. (1988) demonstrated induction of lipid peroxidation in renal tissue of hamsters after the oral administration of Tl malonate. The occurrence of peroxidative processes were referred to depression of GSH level in the kidney. GR activity was not determined by Aoyama et al. (1988). In our experiments lipid peroxidation in renal tissue was not influenced by Tl neither in 10- nor in 55-day-old rats (data not shown). The increased activity of GR (Fig. 9) was reflected by an enhanced GSH/GSSG ratio (data not shown). Probably the enhanced GSH/GSSG ratio prevented induction of peroxidative processes in renal tissue of rats, thereby showing that other reasons were more decisive for the development of the symptoms of Tl nephrotoxicity.

11.4. EFFECTS OF ANTIOXIDANT VITAMINS ON Tl-INDUCED NEPHROTOXICITY

11.4.1. Effect of Vitamin B_2 (Riboflavin)

It has been reviewed in the literature (Saddique and Peterson, 1983; Mulkey and Oehme, 1993) that Tl binds to riboflavin (B_2), and it has been claimed that many features of its toxicity are similar to those found in riboflavin deficiency (Cavanagh, 1991). Moreover, B_2 is known to increase GR activity in several organs, such as renal tissue (Bates, 1993; Appenroth et al. 1996b), and it possesses antioxidant properties preventing lipid peroxidation (Toyosaki, 1992). Therefore we tried to diminish Tl nephrotoxicity by the administration of B_2 as it was done successfully in chromate nephrotoxicity in young and adult rats (Appenroth et al., 1996b). It was not possible to influence oliguria (data not shown) and/or proteinuria (Fig. 10) in 55-day-old rats, neither by concomitant administration of Tl and B_2 nor by B_2 administration 3 h after Tl administration. This interval was shown to be protective in chromate nephrotoxicity in 10- and 55-day-old rats (Appenroth et al., 1996b). Because of the lack of a protective effect of B_2 in adult rats, we did test it in young rats. Since the administration of a pharmacological B_2

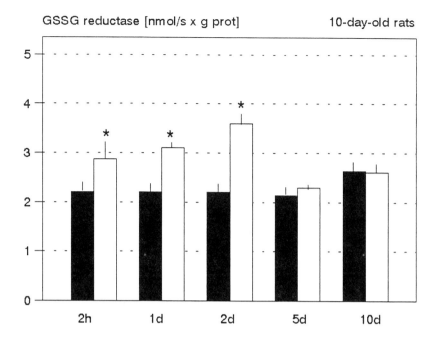

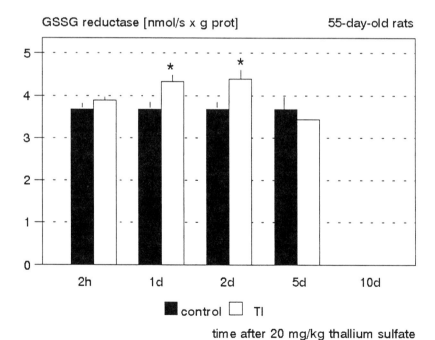

Figure 9. Effect of one single dose of thallium sulfate on the activity of GSSG reductase (GR) in renal tissue of 10- and 55-day-old at different times after thallium administration in comparison to untreated control rats. *, statistically significant differences between control- and Tl-treated rats ($n = 6$, $p \leq 0.05$).

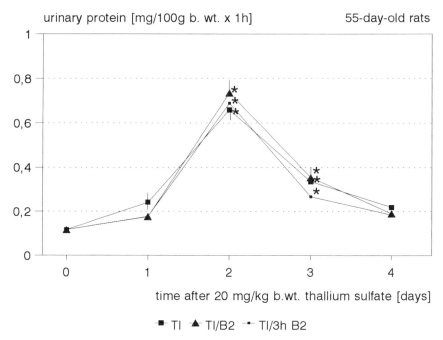

Figure 10. Effect of vitamin B$_2$ (B$_2$) on Tl-induced proteinuria in 55-day-old rats at different times after the administration of one single dose of thallium sulfate. 50 mg/kg b.wt. riboflavin were administered concomitantly with thallium sulfate (Tl/B$_2$) or 3 h after Tl (Tl/3h B$_2$). Protein excretion at time 0 (before the administration of Tl) were used as control. *, statistically significant differences in comparison to control value (time 0) ($n = 6$, $p \leq 0.05$).

dose did not influence Tl nephrotoxicity, we conclude that it is unlikely that at least the nephrotoxic effects of Tl can be due to decreased levels of riboflavin.

11.4.2.   Effect of Vitamin E (α-Tocopherol)

Vitamin E is located in cellular membranes and its primary role is the prevention of lipid peroxidation (Chow, 1991). In rats a small part of vitamin E metabolites is excreted via urine (Drevon, 1991). Since Tl toxicity in hamsters was reported to be associated with increased lipid peroxidation (Aoyama et al., 1988), we wanted to test whether or not vitamin E was able to prevent Tl nephrotoxicity in young and adult rats. Figure 11 clearly shows the protective effect of vitamin E on Tl-induced proteinuria in 10- and 55-day-old rats. But, the reason for this is not to explain with the antioxidant effect of vitamin E because (1) in our rats no increase in the lipid peroxidation as consequence of Tl administration was detected in renal tissue (data not shown), and (2) Tl concentration was decreased in the renal medulla of vitamin-E-treated rats

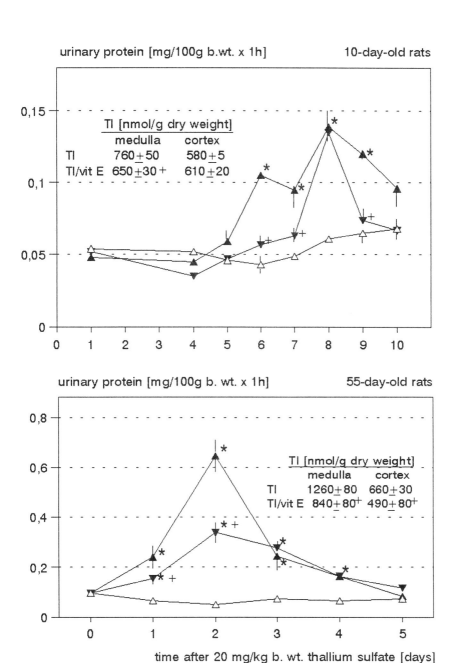

Figure 11. Effect of vitamin E on Tl concentration in renal medulla and cortex (at the eighth day in 10-day-old rats, at the second day after Tl in 55-day-old rats) as well as on Tl-induced proteinuria in 10- and 55-day-old rats at different times after Tl. 1 g α-tocopherolacetate/kg b.wt was administered intramuscularly 12 h before thallium sulfate (Tl/vit E). *, statistically significant differences in comparison to control ($n = 6$, $p \leq 0.05$); +, statistically significant differences between Tl and Tl/vit E ($n = 6$, $p \leq 0.05$).

(Fig. 11). No evidence exists in the literature that vitamin E is able to decrease the concentration of Tl, other metals, or any foreign substances stimulating its excretion. From our results we conclude that there is a protective effect of vitamin E against Tl nephrotoxicity in 10- as well as 55-day-old rats by decreasing Tl concentration in renal medulla, but the mechanism remains completely unknown.

11.5. SUMMARY AND CONCLUSIONS

After one single dose Tl was accumulated in renal tissue (approximately 100-fold in comparison to serum concentrations) preferentially in the medulla. Thallium concentrations were significantly lower in the renal tissue of young rats.

After the administration of 20 mg Tl_2SO_4/kg b.wt. the body weight of young and adult rats did not change in comparison to body weight gain of control rats. All effects on renal morphology and function as well as the effects on Na^+/K^+-ATPase and GR went back to control values within the experimental period of 10 days.

In adult rats after 2 days one single Tl administration caused severe destruction of TALs, which is confirmed by the reduced diuretic effect of furosemide. The structure of glomeruli and proximal tubules remained unaffected. Morphologic destructions regenerated completely up to the tenth day after Tl administration. In young rats renal structure remained completely unaffected. However, the restricted efficiency of furosemide indicated a functional lesion as occurred in TALs of adult rats.

Functional disturbances were characterized in adult rats by oliguria, decreased GFR and increased BUN, and proteinuria. All these effects were less or not (in case of GFR and BUN) expressed in young rats. Excretion of Na and K was diminished in adult but enhanced in young rats. Lipid peroxidation was not influenced by Tl in young and adult rats. The lower nephrotoxicity of Tl in young rats was caused by a lower accumulation of Tl in renal tissue. In addition, more rapid regeneration processes and a higher activation of GR were discussed to protect the kidney of young rats against toxic Tl effects.

Our investigations only in part supported the hypotheses on the mechanisms of Tl toxicity under discussion:

1. Thallium impairs cell function by inhibiting (high levels of Tl) or activating (low levels of Tl) of Na^+/K^+-ATPase activity. We conclude that the activity of ATPase is an indirect factor in Tl nephrotoxicity by influencing Tl concentration in renal tissue.

2. Thallium causes a deficiency of riboflavin and riboflavin-derived cofactors. We conclude that Tl does not act as flavoprotein inhibitor, for example, in the case of GR. Vitamin B_2 had no protective effect.

Vitamin E exerts its protective effect in the Tl-induced nephrotoxicity of young and adult rats by decreasing the Tl concentration in renal tissue and not by its known antioxidant function.

REFERENCES

Aoyama, H., Yoshida, H., and Yamamura, Y. (1986). Acute poisoning by intentional ingestion of thallous malonate. *Hum. Toxicol.*, **5**, 389–392.

Aoyama, H., Yoshida, M., and Yamamura, Y. (1988). Induction of lipid peroxidation in tissues of thallous malonate-treated hamsters. *Toxicology*, **53**, 11–18.

Aperia, A., and Elinder, G. (1981). Distal tubular sodium reabsorption in the developing rat kidney. *Am. J. Physiol.*, **240**, F487–F491.

Appenroth, D., and Bräunlich, H. (1984). Age differences in cisplatinum nephrotoxicity. *Toxicology*, **32**, 77–82.

Appenroth, D., and Bräunlich, H. (1986). Age-dependent qualitative and quantitative changes in physiological proteinuria in rats. *Z. Versuchstierk.*, **28**, 77–82.

Appenroth, D., and Bräunlich, H. (1988). Age dependent differences in sodium dichromate nephrotoxicity in rats. *Exp. Pathol.*, **33**, 179–185.

Appenroth, D., Gambaryan, S., Gerhardt, S., Kersten, L., and Bräunlich, H. (1990). Age dependent differences in the functional and morphological impairment of kidney following cisplatin administration. *Exp. Pathol.*, **38**, 231–239.

Appenroth, D., Gambaryan, S., Winnefeld, K., Leiterer, M., Fleck, C., and Bräunlich, H. (1995). Functional and morphological aspects of thallium-induced nephrotoxicity in rats. *Toxicology*, **96**, 203–215.

Appenroth, D., Tiller, S., Gambaryan, S., Winnefeld, K., Fleck, C., and Bräunlich, H. (1996a). Ontogenetic aspects of thallium-induced nephrotoxicity in rats. *J. Appl. Toxicol.*, **16**, 235–243.

Appenroth, D., Schulz, O., and Winnefeld, K. (1996b). Riboflavin can decrease the nephrotoxic effect of chromate in young and adult rats. *Toxicology*, **87**, 47–52.

Appenroth, D., and Winnefeld, K. (1998). Reconsideration of some hypotheses on the mechanism of toxic thallium action with special regard to nephrotoxicity in rats. *Arch. Toxicol.*, in press.

Bates, C. (1993). Riboflavin. *Int. J. Vit. Nutr. Res.*, **63**, 274–277.

Bernard, A., Robert, R., and Lauwerys, D. (1991). Proteinuria: Changes and mechanisms in toxic nephropathies. *Toxicology*, **21**, 373–405.

Bräunlich, H., Pfeifer, R., Grau, P., and Reznik, L. (1989). Renal effects of vanadate in rats of different ages. *Biomed. Biochem. Acta*, **48**, 569–575.

Britten, J. S., and Blank, S. (1968). Thallium activation of the (Na^+-K^+)-activated ATPase of rabbit kidney. *Biochem. Biophys. Acta*, **159**, 160–166.

Bugarin, M. G., Casas, J. S., Sordo, J., and Filella, M. (1989). Thallium(I) interactions in biological fluids: A potentiometric investigation of thallium(I) complex equilibria with some sulfur-containing amino acids. *J. Inorg. Biochem.*, **35**, 95–105.

Careaga-Olivares, J., and Morales-Aguilera, A. (1990). Acute effects of administration of potassium on the organ uptake Tl^+ in the rabbit. *Arch. Invest. Med. Mex.*, **21**, 273–277.

Cavanagh, J. B. (1991). What have we learned from Graham Frederick Young? Reflections on the mechanism of thallium neurotoxicity. *Neuropathol. Appl. Neurobiol.*, **17**, 3–9.

Cavanagh, J. B., Fuller, N. H., Johnson, H. R. M., and Rudge, P. (1974). The effects of thallium salts, with particular references to the nervous system changes. *Quart. J. Med., New Series*, **43**, 293–319.

Celsi, G., Jakobsson, B., and Aperia, A. (1986). Influence of age on compensatory renal growth in rats. *Pediatr. Res.,* **20,** 347–350.

Chandler, H. A., and Scott, M. (1986). A review of thallium toxicology. *J. R. Nav. Med. Serv.,* **72,** 75–59.

Chow, C. K. (1991). Vitamin E and oxidative stress. *Free Rad. Biol. Med.,* **11,** 215–232.

Danilewicz, M., Danilewicz, M., and Kurnatowski, A. (1979). Histopathological changes in the kidney of experimental animals caused by lethal doses of thallium. *Med. Pr.,* **30,** 257–265.

Deneke, S. M., and Fanburg, B. L. (1989). Regulation of cellular glutathione. *Am. J. Physiol.,* **257,** L163–L173.

Dillard, C. J., Hu, M.-L., and Tappel, A. L. (1987). Effect of aurothioglucose on glutathione and glutathione-metabolizing and related enzymes in rat liver and kidney. *Chem.-Biol. Interactions,* **64,** 103–114.

Douglas, K. T., Bunni, M. A., and Baindur, S. R. (1990). Thallium in biochemistry. *Int. J. Biochem.,* **22,** 429–438.

Drevon, C. A. (1991). Absorption, transport and metabolism of vitamin E. *Free Rad. Res. Comms.,* **14,** 229–246.

Gehring, P. J., and Hammond, P. B. (1967). The interrelationship between thallium and potassium in animals. *J. Pharmacol. Exp. Ther.,* **155,** 187–201.

Gilliland, G. L. (1993). Glutathione proteins. *Curr. Opinion Struct. Biol.,* **3,** 875–884.

Herman, M. M., and Bensch, K. D. (1967). Light and electron microscope studies of acute and chronic thallium intoxication in rats. *Toxicol. Appl. Pharmacol.,* **10,** 199–222.

Hologgitas, J., Ullucci, P., Driscoll, J., Grauerholz, J., and Martin, H. (1980). Thallium elimination kinetics in acute thallotoxicosis. *J. Anal. Toxicol.,* **4,** 68–73.

Kazantzis, G. (1986) Thallium. In *Handbook on the Toxicology of Metals,* Friberg, L., Nordberg, G. F., and Vouk, V. B. (eds.). Elsevier/North-Holland Biomed. Press, Amsterdam, New York, Oxford, pp. 549–567.

Kelner, M. J. (1994). Thallium. In *Handbook on Metals in Clinical and Analytical Chemistry,* Seiler, H. G., Sigel, A., and Sigel, H. (eds.). Marcel Dekker, New York, pp. 601–610.

Kersten, L., and Bräunlich, H. (1968). Biologische Normalwerte bei Wistarratten (Jena) verschiedenen Alters. V. Die renale Wasser- und Elektrolytausscheidung. *Z. Versuchstierk.,* **10,** 195–204.

Lehmann, P. A., and Favari, L. (1985). Acute thallium intoxication: Kinetic study of the relative efficacy of several antidotal treatments in rats. *Arch. Toxicol.,* **57,** 56–60.

Leloux, M. S., Nguyen, P. L., and Claude, J. R. (1987). Experimental studies on thallium toxicity in rats. I. Localization and elimination of thallium after oral acute and subacute intoxication. *J. Toxicol. Clin. Exp.,* **7,** 247–257.

Lie, R., Thomas, R. G., and Scott, J. K. (1960). The distribution and excretion of thallium-204 in the rats, with suggested MPC's and a bio-assay procedure. *Hlth. Phys.,* **2,** 334–340.

Lund, A. (1956). Distribution of thallium in the organism and its elimination. *Acta Pharmacol. Toxicol.,* **12,** 251–259.

Melnick, R. L., Monti, L. G., and Motzkin, S. M. (1976). Uncoupling of mitochondrial oxidative phosphorylation by thallium. *Biochem. Biophys. Res. Commun.,* **69,** 68–73.

Moeschlin, S. (1980). Thallium poisoning. *Clin. Toxicol.,* **17,** 133–146.

Mulkey, J. P., and Oehme, F. W. (1993). A review of thallium toxicity. *Vet. Human Toxicol.,* **35,** 445–453.

Sabbioni, E., and Manzo, L. (1979). Metabolism and toxicity of thallium. *Adv. Neurotoxic.,* **1979/1980,** 249–270.

Saddique, A., and Peterson, C. D. (1983). Thallium poisoning. *Vet. Hum. Toxicol.,* **25,** 16–22.

Shabalina, L. P., and Spiridonova, V. S. (1979). Thallium as an industrial poison: Review of literature. *J. Hyg. Epidemiol. Microbiol. Immunol.*, **231**, 247–256.

Thyresson, N. (1950). Experimental investigation on thallium poisoning. Influence of thallium on tissue metabolism. *Acta Dermato-Venerol.*, **30**, 417–441.

Toyosaki, T. (1992). Antioxidant effect of riboflavin in enzymatic lipidperoxidation. *J. Am. Chem. Soc.*, **40**, 1727–1730.

WHO (1973). Safe use of pestizides. 20th report of WHO expert committee on insecticides. *WHO Techn. Refer. Serv.*, **513**, 40–44.

WHO (1996). Thallium. Environmental health criteria 182. World Health Organization, Geneva.

12

THALLIUM TRANSPORT IN CELLULAR MEMBRANES

Tom Brismar

Department of Clinical Neurophysiology, Karolinska Hospital, 17176 Stockholm, Sweden

Thallium in the Environment, Edited by Jerome O. Nriagu.
ISBN 0-471-17755-5. © 1998 John Wiley & Sons, Inc.

12.1. INTRODUCTION

The Tl^+ ion is transported in the cell membrane in the same fashion as K^+, although there are quantitative differences. Since K^+ is the major intracellular cation with crucial importance for a number of identified cellular functions, great efforts have been made in the exploration of the cellular K^+ transport. Thallium-204, ^{201}Tl, and ^{86}Rb have more convenient properties as radioactive tracers than ^{42}K and have therefore been used to elucidate the mechanism of the K^+ transport. Differences in ionic radii between K^+, Tl^+ and other monovalent cations have been utilized in studies of the selectivity filter of different cation channels. Three major routes for K^+ transport have been identified: ion channels, cationic pumps, and coupled transport. In the following the different pathways for K^+ and Tl^+ in the membrane will first be described separately; thereafter a description is given of their relative contributions to the Tl^+ uptake in different cells.

12.2. THALLIUM TRANSPORT IN ION CHANNELS

12.2.1. K^+ Channels and the Resting Potential

The K^+ channels in the cell membrane are characterized by their high selectivity for K^+ and some other cations such as Tl^+ and Rb^+ allowing the "passive" outward flow of K^+. This results in an excess of negative charges and a negative potential inside the cell, which is described by the Nernst equation $E_K =$ const. $\ln([K]_o[K]_i)$ where E_K is the equilibrium potential for K^+, const. is 0.025 V, and [ ] denotes outside and inside K^+ concentrations, which in mammals are about 5 and 150 mM, respectively. Solving the equation with these values results in $E_K = -85$ mV. E_K has a dominating influence on the cellular resting potential, which amounts to -50 to -90 mV in different cell types, since the K^+ conductance is dominating in the membrane. Hagiwara et al. (1972) calculated the permeability of a number of monovalent cations from their depolarizing "efficiency" in the resting squid axon. They found the permeability ratios to be 0.18, 0.25, 0.71, 1.82, and 1.0 for Cs^+, NH_4^+, Rb^+, Tl^+, and K^+, respectively. This indicated that Tl^+ was even more permeant in the membrane than its natural congener. A description of the Tl^+ permeability of different voltage-dependent (gated) K^+ channels follows. Except for the inwardly rectifying K^+ channel, these channels are activated by depolarization of the membrane causing an outward K^+ current, since the electrochemical driving force for K^+ is outward. All K^+ channels are highly permeable to Tl^+.

12.2.2. Delayed Rectifier K^+ Channels

Several different cations have been compared for their permeability ratios in the K^+ channel of the nodal axon membrane of frog myelinated nerve fibers (Hille, 1973). The membrane permeability properties were analyzed with voltage clamp, which makes it possible to record the current under membrane potential control. The channel was first activated (opened) by a depolarization, then pulses of different amplitudes were applied causing changes in the electrochemical driving force on the permeant ion. In this way the reversal potential was determined, which is the potential where the net ion current was zero, balancing between outward and inward. The permeability ratio was calculated from the change in reversal potential (ΔE_r) using the relation $\Delta E_r =$ const. $\ln(P_x[x]_o/P_K [K]_i)$ where P_x is the permeability and $[x]_o$ is the concentration of the tested ion in the external solution. Obviously, E_r equals zero when K^+ is the cation in the external solution. Equimolar substitution with a test ion such as Tl^+ instead of K^+ causes a shift in the reversal potential by 18.4 mV, which was equivalent to a permeability ratio of 2.3. Hille (1973) found a relationship between the permeability ratio and the size of different ions suggesting that the selectivity of the ionic channel to a certain extent can be explained on geometrical grounds. A pore diameter of 3.0–3.3 Å would allow K^+, Tl^+, and Rb^+ to pass and exclude ions with an ionic radius larger than ammonium.

The Na^+ and Li^+ ions have very low permeability ratios in spite of their small ionic radii. This has been explained by the hydration of the cations and the presence of polarized sites (or binding sites) in the ion channel (see Hille, 1992, 1973). The energy required to move cations from water to the binding site would be larger for small cations such as Na^+ and Li^+ than for K^+. However, if the binding site has a strong electrostatic field, the attraction of small cations would be larger, which would change the sequence of the permeability ratios.

Molluscan neurons have several types of channels for the outward K^+ currents, one being the delayed rectifier, which was studied for its ion selectivity properties by Reuter and Stevens (1980). They found that an equimolar change from K^+ to Tl^+ as the main cation of the external solution shifted the reversal potential by -6.2 ± 2.0 mV, which corresponds to a permeability ratio P_{Tl}/P_K of 1.29. When the cells were exposed to solutions with high $[Tl^+]$ on the outside and high $[K^+]$ on the inside, the instantaneous current–voltage relationship revealed a conductance ratio g_{Tl}/g_K of 1.30. Similar to the findings in the myelinated nerve, the calculated values of P_{Tl} and g_{Tl} are higher than expected from the size of Tl^+, which lies between that of K^+ and Rb^+. This deviation may be due to a more favorable interaction energy of Tl$^+$ with some constituent of the channel (Reuter and Stevens, 1980).

12.2.3. Inwardly Rectifying K$^+$ Channels

Figure 1 shows the current fluctuations in a small membrane patch in a human glioma cell when the potential was clamped to different levels. Hyperpolarization of the membrane patch resulted in steps of inward current. This finding was in agreement with the properties of the inwardly rectifying K$^+$ channel. From the measured size of the step changes in the current fluctuation, the single channel conductance was estimated to be 27 pS. Figure 2 shows the current recorded from the whole cell in association with membrane potential changes. With 5 mM K$^+$ in the extracellular solution, there is a substantial inward current when the membrane is hyperpolarized. Depolarization of the membrane has no effect. Substitution with 5 mM Tl$^+$ for K$^+$ results in larger inward current, which demonstrates that the channel for this current is more

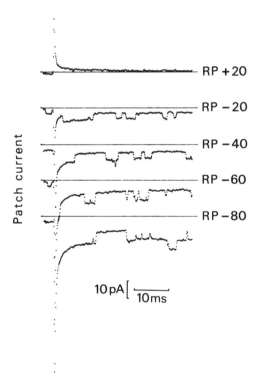

Figure 1. Inward current in cell-attached patch of human glioma cell membrane. Membrane potential change from resting potential (RP) indicated on each record. Pipette solution on membrane outside contained 160 mM K$^+$, which resulted in large steps of inward current. From Brismar and Collins (1989).

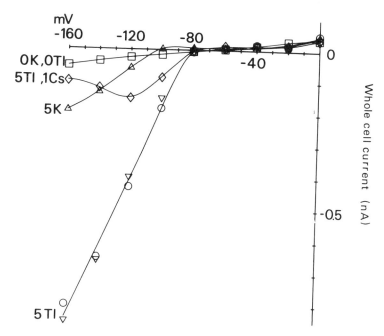

Figure 2. Inward current of Tl$^+$ and the effect of Cs$^+$ in human glioma cells. The currents were five times larger in 5.0 mM Tl-acetate than in 5.0 mM K-methylsulfate (both were Cl$^-$-free solutions). Addition of 1.0 mM CsCl blocked the inward Tl$^+$ current in a voltage-dependent manner. From Brismar and Collins (1989).

permeable for Tl$^+$ than K$^+$. Estimated in this way Tl$^+$ was 3.5 times more permeant than K$^+$. Addition of 1 mM Cs$^+$ effectively blocks the current (Brismar and Collins 1989).

Inwardly rectifying K$^+$ channels have been detected in different types of cells such as muscle, glia, and egg cells. This type of channel is to some extent open at resting potential and it differs from other K$^+$ channels in that its conductance decreases when the membrane is depolarized. The function is probably to stabilize the resting potential and yet not counteract the depolarization once it has occured. This may be economic in cells with long-lasting depolarizations such as heart and egg cells (see Hille, 1992). In glial cells it has been speculated that inwardly rectifying K$^+$ channels may facilitate uptake of K$^+$, but this has not been verified experimentally (Brismar and Collins 1989; Brismar et al., 1995b).

The cation selectivity of the inwardly rectifying channel has been determined in egg cell membranes of the starfish (Hagiwara and Takahashi, 1974). Changes in the concentration of Tl$^+$, K$^+$, and Rb$^+$ in Tris media revealed a Nernstian slope of the curve relating membrane potential to the ionic concentration. Tl$^+$ was more effective than K$^+$, a difference which was equivalent to a P_{Tl}/P_K of 1.5. The ionic selectivity was also calculated from the change in the chord conductance as measured from the effect of current pulses on the membrane potential in different ionic solutions, which gave a permeability ratio P_{Tl}/P_K of 1.3–1.5. A remarkable finding in this study was that the conductance decreased when the cells were bathed in mixtures of Tl^+ and K^+. This was named *anomalous mole fraction dependence* and was further investigated by Hagiwara et al. (1977). They found that the conductance went through a minimum (ca. 40% of P_K in pure K^+ media) when the external solution contained about 25% Tl^+ and 75% K^+ (Hagiwara et al., 1977).

Another effect of high $[Tl^+]$ was seen if frog sartorius muscle, where the inwardly rectifying current rapidly inactivates when the external solution contains 80 mM Tl^+, which may be due to a conformational change in the channel caused by Tl^+ (Stanfield et al., 1981). The ionic specificity of the inwardly rectifying channel was similar to that seen in the egg cell membrane. When the external solution was changed from 80 mM K^+ to 80 mM Tl^+, this shifted the reversal potential for the current by 12 mV indicating a ratio P_{Tl}/P_K of 1.66. However, in mixtures of Tl^+ and K^+ the inward currents were less than in pure K^+ and Tl^+ solutions, that is, the conductance exhibited anomalous mole fraction dependence (Ashcroft and Stanfield, 1983).

12.2.4. Ca^{2+}-Activated K^+ Channels

Big conductance Ca^{2+}-activated K^+ (BK) channels have been found in most types of cells. Because of their high conductance (100–250 pS) the single-channel currents are prominent and easily identified in patch clamp studies. The selectivity of the Ca^{2+}-activated K^+ channel has been studied in cultured skeletal muscle of the rat (Blatz and Magleby, 1984). A permeability ratio P_{Tl}/P_K of 1.2 was determined from the measured changes in the reversal potential of the single channel current (see above) obtained at different test pulses with the membrane patch bathed in K^+ on the outside and K^+ or Tl^+ on the inside. The outward current carried by Tl^+ was smaller than expected from the permeability ratio, suggesting that outward flow of Tl^+ is impeded when K^+ is present on the outside of the membrane. In these calculations the incomplete ionization (85%) of $TlNO_3$ was accounted for. When this is done, the data of Gorman et al. (1982) indicate that P_{Tl}/P_K is about 1.2 in the Ca^{2+}-activated K^+ channel in Aplysia neurons and the light-dependent K^+ channels of photoreceptors of the scallop.

The single-channel conductance of skeletal muscle K^+ channels have been investigated in experiments where membrane vesicles were inserted in artificial bilayers. The sarcoplasmic reticulum (SR) cation channel of rabbit muscle has a maximum limiting conductance of 223 pS in symmetrical single-salt solutions of K^+ (Bell and Miller, 1983) and 57.5 pS in pure Tl^+ acetate solutions (Fox and Ciani, 1985). In rat skeletal muscle big K^+ channels inserted in lipid bilayers the conductance for Tl^+ (in single-ion solutions) was 130 pS or 0.41 relative to that of K^+. However, under bi-ionic conditions with K^+ outside and Tl^+ inside, the effect on the reversal potential, calculated from $P_{Tl}/P_K =$

$\exp(-FV/RT)$ indicated that $P_{\text{Tl}}/P_{\text{K}}$ was 1.3. In mixtures of Tl^+ and K^+ the conductance exhibited a minimum that could be due to multiple occupancy of permeable ions in the channel (Eisenman et al., 1986). Thus Tl^+, K^+, Rb^+, and Cs^+ are all permeable in the BK channel, but they have different effect on the gating, which might be explained by their binding to different sites in the pore. Presence of Tl^+ in the external solution reduced the size of the single-channel current. This is because Tl^+ has higher affinity than K^+ for a binding site in the pore. This binding site may be located near the outer opening of the channel. Cs^+ or Rb^+ in the external solution affect gating because they may have high affinity to an inner binding site that is not affected by Tl^+ or K^+ in the external solution (Demo and Yellen, 1992).

Other types of Ca^{2+}-activated K^+ channels are the small conductance (SK) channel and some other K^+ channels with intermediary conductance levels (see Blatz and Magleby, 1987). The SK channel accounts for the long-lasting afterhyperpolarization in cultured skeletal muscle. Apamin, which is the neurotoxin in bee venom, blocks the SK channel and the afterhyperpolarization, and neither is affected by tetraethylammonium (TEA) which is a blocker of the BK Ca^{2+}-activated K^+ channel. The SK channel has a single channel conductance of only 10–14 pS (Blatz and Magleby, 1986).

12.2.5.   A-Current K^+ Channels

The A-current is a transient outward K^+ current that is important for the repolarization of the membrane in excitable cells. It has been detected in many cell types such as neurons of vertebrates and invertebrates, skeletal and cardiac muscle, oocytes, and endocrine cells (Rogawski, 1985). The A-current seems particularly important for the regulatation of the spike frequency during repetitive firing. This channel differs from that of the delayed outward current and the Ca^{2+}-activated channel in its rapid inactivation. The gating of the channel can be described by two independent time-requiring processes, activation and inactivation, which both are controlled by voltage (Connor and Stevens, 1971). During an action potential, the channel is activated and the resulting outward A-current helps to bring back the membrane potential to its resting level. However, the channel is rapidly inactivated by the depolarization, and the channel will therefore not counteract the subsequent depolarization of the membrane when the next action potential is initiated. The selectivity of the A-current channel has been determined from the effect of various cations on the reversal potential of the current. The channel is permeable to Tl^+, K^+, Rb^+, NH_4^+, and Cs^+ with relative permeabilities 2.04, 1.0, 0.73, 0.18, and 0.14, respectively (Taylor, 1987).

12.2.6.   Other K^+ Channels

The high permeability of Tl^+ in the K^+ channels is a feature shared by all studied K^+ channels. Rod photoreceptors have a channel that is activated by hyperpolarization of the membrane with low selectivity between Na^+ and K^+, the ratio P_{Na}/P_K being 0.33. The ratio P_{Tl}/P_K of >1.55 was calculated from changes in reversal potential under bi-ionic conditions. When maximum slope conductances for Tl^+ and K^+ were compared the ratio g_{Tl}/g_K is 1.07 (Wollmuth and Hille, 1992).

The K^+ channels in the kidney are important for the maintenance of the cell-negative membrane potential. One type of renal K^+ channel designated ROMK2 has been expressed in oocytes and studied with patch clamp (Chepilko et al., 1995). The reversal potential indicated that P_{Tl}/P_K is 1.56 and the single channel conductances in isotonic Tl^+ and K^+ solutions are 25 and 35 pS, respectively, that is, g_{Tl}/g_K is 0.7. Presence of Tl^+ made the channel more flickery, the longer closed states were abolished, and the time constant of the open state was reduced from 19 to 4 ms.

12.2.7. Neuronal Na^+ Channels and Cholinergic Channels in the Muscle Endplate

The Na^+ channels of the nodal membrane in myelinated nerve fibers have been studied in great detail with regard to their gating mechanism and ionic selectivity properties. In frog nerve the permeability sequence is $Na^+ = Li^+ > Tl^+ > K^+$, where the ratio P_{Tl}/P_{Na} is 0.33. High concentrations of Tl^+ were found to be directly toxic to the nodal membrane, which became low-resistant with irreversible damage to the function of the Na^+ channels (Hille, 1972).

Activation of the endplate channel in skeletal muscle was studied from the currents elicited by brief iontophoretic puffs of acetylcholine to the endplate (Adams et al., 1980). Calculation of the selectivity was made from the shift in the reversal potential, showing a permeability ratio P_{Tl}/P_{Na} of 2.51 and P_K/P_{Na} of 1.11. Other alkali metal ions showed a permeability in this channel with a sequence similar to that of their mobilities in aqueous solutions. This means that the selectivity of this channel is very weak, except for Tl^+, which had much higher permeability than predicted from the mobility ratio Tl^+/Na^+ of 1.49.

12.3. CATION PUMPS

12.3.1. Na,K-ATPase Activation

The high intracellular K^+ concentration is maintained by the ATP requiring "active" transport of K^+ into the cell. Under physiological conditions the enzyme Na^+,K^+-ATPase transports $3Na^+$ out of the cell and $2K^+$ into the cell for each ATP molecule that is hydrolyzed. Presence of cytoplasmic Na^+ was long thought to be necessary for the Na^+, K^+-ATPase activity, however, Na^+ may be replaced by H^+ or Li^+ on the intracellular side and K^+ can be replaced by Tl^+, Rb^+, Cs^+, Li^+, and HN_4^+ on the extracellular side (see Karlish, 1989). Of these K^+ congeners Tl^+ is the most effective one in activating the enzyme, even more so than K^+.

Britten and Blank (1968) found the affinity of rabbit kidney Na^+,K^+-ATPase to be approximately 10 times greater for Tl^+ than for K^+ on the K^+ activating site. Comparison of different cations for their capacity to activate Na,K-ATPase in microsomal preparations of rat brain showed that the half-maximum activating concentration (K_m) for Tl^+, K^+, Rb^+, HN_4^+, and Cs^+ was 0.15, 0.80, 0.74, 6.1, and 6.2 mM, respectively (Robinson 1975). These cations had a similar order of potency in preventing the blockade of the Na^+/K^+ pump ("late effects") in mammalian cardiac muscle caused by K^+-free solutions (Eisner and Lederer, 1979). Activation of the Na^+,K^+-ATPase by Tl^+ as compared to K^+ was 10 times more effective in rat liver plasma membranes, where the temperature, pH, and Mg^{2+} dependencies of Na^+,K^+- and Na^+,Tl^+-ATPase were similar (Favari and Mourelle, 1985).

The stochiometry of the Na^+/K^+ pump creates a net outward current that can be used to study the pump activity. The pump is stimulated by depolarization of the membrane, which is indicated by an increase in the outward current. This has been demonstrated in several cell types, and is considered to be related to the liberation of Na^+ at the external pump site (Apell, 1989). The Na^+/K^+ pump is stimulated by Na^+-free external solution due to a competition between Na^+ and K^+ for a common binding site as indicated by a lower K_m in Na^+-free media (Post et al., 1960).

Omay and Schwarz (1992) studied the stimulating effect of various external cations on the endogenous Na^+/K^+ pump current (I_p) in *Xenopus* oocytes. In Na^+-free solutions, the activation by external cations was inhibited by depolarization of the membrane. The voltage dependency of the half-maximum cation concentration for I_p activation (K_m) could be described by a single exponential function (Boltzmann distribution of electronic charges), suggesting that activation is associated with the movement of an effective charge of 0.4 elementary charges. At zero voltage K_m for pump activation was 0.10 mM for Tl^+, 0.63 mM for K^+, 0.71 mM for Rb^+, 9.3 mM for NH_4^+, and 12.6 mM for Cs^+. In mammalian cardiac muscle fibers the current (I_p)–voltage relation of the Na^+ pump was similarly studied in the presence of different cations (Bielen et al., 1991, 1993). In Na^+ containing external solution half maximum I_p occurred at 0.4 mM Tl^+, 1.9 mM K^+, and 5.7 mM NH_4^+. In Na^+-free external solution K_m was 0.05 mM for Tl^+ and 0.08 mM for K^+ at zero membrane potential. The shape of the current–voltage relation was independent of the cation species used for activation. The voltage dependency of K_m suggested the movement of about 0.26 elementary charges in the electrical field across the membrane. The sequence of the activation potency $Tl^+ > K^+ > NH_4^+ > Cs^+$ is similar to the selectivity sequence of the voltage-gated K^+ channels (see above). These observations are in support of the hypothesis that the external K^+ binding site is located in an "ionic well" located within the electric field of the membrane, which is accessible via an ionic filter similar to that of the K^+ channels (Stürmer et al., 1991).

Comparison of the Tl^+ uptake in artificial salt solution and in complete medium demonstrated that there are other factors than the ionic environment and the potential that have as strong stimulatory effect on the Na^+/K^+ pump (Brismar et al., 1994). Figure 3 demonstrates a fourfold increase in the ouabain-inhibitable fraction of the steady state ^{201}Tl accumulation when Jurkat cells were incubated in complete medium as compared with Ringer's solution.

12.3.2. H,K-ATPase

Another means for active K^+ uptake in cells is through the action of the H^+,K^+-ATPase, which is present not only in the parietal cells of the gastric mucosa where it is involved in acid secretion, but also in the epithelial cells of the distal colon and in the collecting ducts of the kidneys (see van Driel and Callaghan, 1995). The gastric H^+/K^+ exchange mechanism is electro-neutral with a stochiometry of $2H^+ : 2K^+ : 1ATP$ at pH 6.1, which falls to $1H^+ : 1K^+ : 1ATP$ at a luminal pH < 3 (see Rabon and Reuben, 1990). The gastric H^+,K^+-ATPase belongs to the P-type ATPase family and is closely related to the Na^+,K^+-ATPase with 60% identical amino acid sequence in the catalytic α-subunit.

The steps in the enzymatic reaction is described by changes of the enzyme between two conformations E_1 and E_2 with ion binding sites on the cytoplasmic

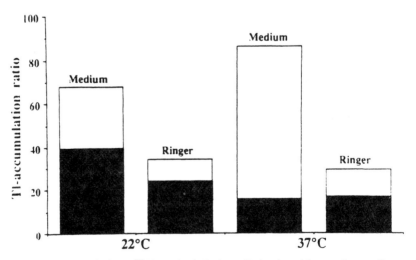

Figure 3. Effect of ouabain on ^{201}Tl uptake in Jurkat cells incubated in complete medium or in Ringer's solution at different temperatures. Cells were incubated for 120 min. Bars show mean values of duplicate samples.

and the luminal side, respectively. The hydronium ion H_3O^+ binds to a cytoplasmic site on the enzyme together with a bound ATP molecule ($E_1H_3O^+$). Transfer of one phosphate from ATP converts the enzyme to E_2–PH_3O^+, which has low affinity for H_3O^+ and high affinity for K^+. Therefore H_3O^+ is liberated and K^+ is bound (E_2–PK^+). The enzyme is then rapidly dephosphorylated to (E_2K^+), which is converted to $E_2(K^+)$ where K^+ is occluded during its transmembrane transport. Binding of ATP to $E_2(K^+)$ promotes conversion to E_1, which releases K^+ into the cytoplasm and binds H_3O^+, completing the reaction cycle (see van Driel and Callaghan, 1995).

The activation by various cations of the enzyme was studied by Forte et al. (1976) in the fundic and the main body of the gastric mucosa in different species (bullfrog, rabbit, pig, and rat). V_{max} was nearly the same for all cations tested, whereas there was a wide variation in the cationic affinity with apparent values of K_m, namely 1.4, 0.25, 0.17, and 0.013 mM for NH_4^+, Rb^+, K^+, and Tl^+, respectively. Thus the potency of Tl^+ in activating the H^+,K^+-ATPase was more than 10 times that of K^+.

A novel type of H^+,K^+-ATPase has been detected in the gastric mucosa that is present in all cell types of gastric glands (Hofer and Machen, 1992). Uptake of K^+ through this mechanism was unaffected by various K^+ transport blockers such as barium, bumetanide, or ouabain. Similar to the property of other H^+,K^+-ATPases in kidney and rat distal colon, this type of gastric cell H^+,K^+-ATPase is blocked by SCH28080 in 100-fold higher concentrations than required to block the H^+,K^+-ATPase of the parietal cells. Both types of H^+,K^+-ATPases are activated by Tl^+ with higher potency than K^+, that is, these are efficient routes for Tl^+ uptake in the gastric mucosa.

Vascular smooth muscle cells seem to have a similar type of H^+,K^+-ATPase-dependent transport mechanism since studies of the Rb^+ uptake in cultured cells (from a cell line and from primary culture) revealed a bumetanide- and ouabain-insensitive component, which was affected by changes in pH and was blocked by specific blockers to this enzyme such as omeprazole, NC-1300-B, and SCH28080. The ^{86}Rb uptake was reduced 19–37% by these blocking agents, which demonstrated that a substantial fraction of the Rb^+ uptake was due to the H^+,K^+-ATPase activity in vascular smooth muscle cells (McCabe and Young, 1992). Changes in Tl^+ uptake has not been investigated in presence/absence of specific H^+,K^+-ATPase blocking agents, however, it seems likely that a similar fraction of the Tl^+ transport into smooth muscle cells is mediated through this K^+/H^+ exchange mechanism.

12.4.   COUPLED TRANSPORT

A very important mechanism for K^+ and Tl^+ transport through the plasma membrane is through the action of the coupled NaCl–KCl transport, which has been identified in numerous animal cell species. This transporter is activated in connection with the regulatory volume increase that follows when cells have been shrunk in a hypertonic medium (see Jensen et al., 1993). The first evidence of Tl^+ transport through the cotransport mechanism was obtained in Ehrlich ascites tumor cells, where incubation in furosemide or Cl^--free medium reduced the initial rate of Tl^+ uptake with about 50%, which was similar to the effect on the Rb^+ uptake (Bakker-Grunwald, 1979). Careful analysis of the coupling of the Na^+, K^+, and Cl^- flux and the stochiometry of the transporter has been carried out in Ehrlich ascites cells (see Geck and Heinz, 1986; Jensen et al., 1993). A change in the concentration of one of these ions in the incubation medium was tightly coupled to a change in the uptake of the other two, with a stochiometry of approximately $1:1:2$ for Na^+, K^+, and Cl^-. This transport is electroneutral and it is inhibitable by furosemide, bumetanide, and other loop diuretics. In fact, the specific action of these inhibitors (and the lack of effect of anion exchange inhibitors such as the stilbene derivatives SITS and DIDS) has been utilized to characterize the cation transport in many studies.

Tl^+, Rb^+, and Cs^+ can more or less replace K^+ in the furosemide-inhibitable ion uptake. The relative rate constants for Tl^+, K^+, and Rb^+ were 6.4, 1.0, and 0.73, respectively, in 6-day old cells, and the figures were similar in cells that were 12 days old (Sessler et al., 1986). The requirement of Na^+ and Cl^- for the ouabain-resistant Tl^+ uptake is illustrated in Figure 4, showing ^{201}Tl uptake in human glioma cells. Virtually no Tl^+ accumulated in Na^+- and Cl^--free solutions.

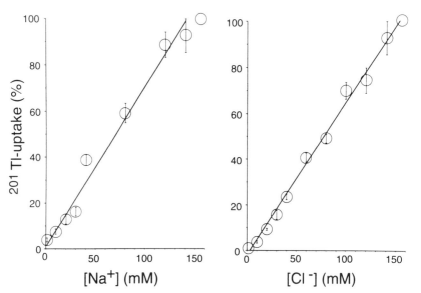

Figure 4. Effect of changes in [Na^+] and [Cl^-] of extracellular solution on rate of Tl^+ uptake. Each symbol represents the mean $\pm$ SEM ($n = 8$). Cells incubated for 10 min at 37°C with ^{201}Tl in presence of 1 mM ouabain. Regression analysis indicated a linear relationship between ionic concentration and ^{201}Tl uptake.

The cotransport is driven by the concentration gradients, which are created by the action of the Na^+/K^+ pump. The transporter is in equilibrium when $[K^+]_0 \times [Na^+]_0 \times [Cl^-]_0^2 = [K^+]_i \times [Na^+]_i \times [Cl^-]_i^2$, which has been verified experimentally in mouse fibroblasts (see Saier and Boyden, 1984). Although the cotransporter functions passively in either direction, its activity may be stimulated by an elevation of cyclic-AMP elicited by adrenergic stimulation (see Chipperfield, 1986; Lytle and Forbush 1992). However, this effect is variable depending on the system studied, for example, in human fibroblasts and lymphocytes agents that increase intracellular cAMP inhibit Na^+-K^+-Cl^- cotransport (Owen and Prastein 1985; Feldman, 1992).

12.5. RELATIVE CONTRIBUTION OF DIFFERENT TRANSPORT MECHANISMS

12.5.1. Glial Cells

In cultured astrocytes the uptake of ^{42}K and ^{86}Rb can be largely blocked by the combination of ouabain and furosemide, which indicates that in these cells the K^+ uptake is due mainly to Na,K-ATPase-dependent transport and Na–K–Cl cotransport. Kimelberg and Frangakis (1985) demonstrated that in primary culture of rat astrocytes 20–25% of the ^{42}K uptake after 30-min incubation was inhibited by furosemide (or bumetanide) and about 50% by ouabain. In the study of Tas et al. (1987) the ouabain-inhibitable fraction was about 70%. In cultured human astrocytes only 30–40% of the ^{86}Rb uptake was inhibited by ouabain, 50–60% depended on Na^+ and Cl^- cotransport and the remainder was due to leak channels (Tas et al., 1986).

The rate of the K^+ and Tl^+ uptake has been studied in cultured human glioma cells incubated in Ringer's solution at 37°C (Brismar et al., 1989, 1995b). Figure 5 shows the effect of ouabain and furosemide on the uptake of ^{201}Tl and ^{42}K measured after 10-min incubation at 37°C. Both ^{201}Tl and ^{42}K were inhibited by about 55% in ouabain and 60% in furosemide (or bumetanide, or Na^+- or Cl^--free medium). However, in solutions containing both ouabain and furosemide (or bumetanide), there was a remaining fraction of 10%, which was resistant to both inhibitors. Ba^{2+}, which causes depolarization and is a blocker of voltage-gated K^+ channels in glial cells (Brismar and Collins, 1993; Reichelt and Pannicke, 1993), had no effect on the rate of Tl^+ uptake. Neither had block of K^+ channels with Cs^+ or bupivacaine any effect (Brismar et al., 1995b).

These findings indicated that passive transport of Tl^+ through K^+ channels does not play a role in Tl^+ uptake in cultured glial cells. This might also have been predicted from an electrochemical point of view, since it is the *outward* leak of K^+ (due to its high intracellular concentration) that creates the membrane potential. However, it is not excluded that in vivo K^+ or Tl^+ can be driven into the cells by their electrochemical gradient, if there is a compartmentization

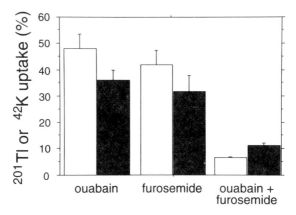

Figure 5. Effect of ouabain (1.0 mM) and furosemide (1.0 mM) on Tl$^+$ and K$^+$ uptake in human glioma cells. Results are means ± SEM of experiments with ^{42}K (filled bars) and with ^{201}Tl (open bars) relative to controls. Cells incubated in Ringer's solution with or without blocking agent for 10 min at 37°C.

between neurons and glial cells. Focal spots of elevated extracellular [K$^+$] may possibly appear in association with sustained neuronal activity.

12.5.2. Ehrlich Ascites Tumor Cells

In many studies it has been assumed that the uptake of ^{201}Tl or ^{86}Rb is representative for the K$^+$ uptake. A quantitative comparison of the rate of cellular uptake of Tl$^+$, Rb$^+$, and K$^+$ was made by Sessler et al. (1986) in Ehrlich mouse ascites cells. They found that in 6-day old cells the ouabain-sensitive and the furosemide-inhibitable fractions were 24 and 64%, respectively, for ^{201}Tl, 34 and 47% for ^{86}Rb, and 32 and 42% for ^{42}K. In presence of ouabain and furosemide there remains a passive Tl$^+$ or Rb$^+$ uptake (Bakker-Grunwald, 1979). This unidentified ionic movement was largest for K$^+$ (26%) and the Na$^+$–K$^+$–Cl$^-$ cotransport (bumetanide–inhibitable fraction) was largest for Tl$^+$. This demonstrated a greater similarity between Rb$^+$ and K$^+$, whereas for Tl$^+$ the Na$^+$–K$^+$–Cl$^-$ cotransport was larger (Sessler et al., 1986).

12.5.3. Blood Cells

Erythrocytes have two components of Tl$^+$ influx: a saturable ouabain-sensitive influx and an ouabain-insensitive influx that is not saturated in high external [K$^+$] or [Tl$^+$] (Gehring and Hammond, 1964; Skulskii et al., 1978). Substitution with Mg(NO$_3$)$_2$ for NaCl and KCl in the extracellular medium maximized the ouabain-sensitive Tl$^+$ transport, which was 0.45 mmol/(L cells) h with K_m as low as 0.025 mM. Transport of Tl$^+$ via the Na$^+$–K$^+$(Tl$^+$)–Cl$^-$ cotransport was inhibited in the Na$^+$- and Cl$^-$-free solution. The kinetics for the remaining "passive" cation transport was considerably more rapid for Tl$^+$ than for Rb$^+$. The inward and outward rate constants were 0.071 and 0.045 min^{-1}, respectively, for Tl$^+$ in human erythrocytes and about twice as rapid in erythrocytes of the rat (Skulskii et al., 1990).

In cultured human lymphocytes (Jurkat cells and EBV-transformed B-cells) the ouabain-sensitive fraction of the Tl$^+$ uptake measured at near steady state (60 min) was 81% in complete culture medium at 37°C (Brismar et al., 1994). Substitution with Ringer's solution for culture medium decreased the ouabain-sensitive fraction to about 50%. The remainder of the uptake in the presence of ouabain was carried through Na$^+$–K$^+$–Cl$^-$ cotransport since it was almost completely inhibited by furosemide (or Na$^+$- or Cl$^-$-free solutions). The rate constants were determined from reciprocal plots of the effect of [Tl$^+$] on the ^{201}Tl uptake at 22°C in K$^+$-free Ringer's solution (Fig. 6). For the ouabain-sensitive uptake K_m is 0.10 mM and the maximum rate (V_{max}) is 12 mmol/(L cells) h, and for the ouabain-insensitive uptake the corresponding figures are 0.16 mM and 31 mmol/(L cells) h.

In a study of the mechanism of multidrug resistance, the properties of cells from a human leukemic cell line (K562) were compared with those of a vincristine-resistant subline (Brismar et al., 1995a). Inhibition of the Na$^+$,K$^+$-

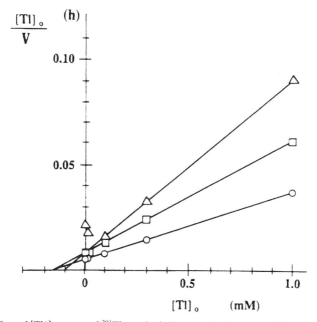

Figure 6. Effect of [Tl$^+$] on rate of ^{201}Tl uptake in Jurkat cells. The rate (V) was measured from ^{201}Tl uptake after 30 min incubation at 22°C in K$^+$-free Ringer's solution without (○) and with ouabain (□). Ouabain-sensitive uptake indicated (△). Data presented in reciprocal plot where [Tl]$_0$/V = K_m/V_{max} + [Tl]$_0$/V_{max}. In the range of 0.1–1.0 mM Tl$^+$ the data were well fitted by straight lines, where the x-intercept is K_m and the slope is V_{max} (maximim rate of Tl$^+$ uptake).

ATPase activity with ouabain decreased ^{201}Tl uptake by 78% in both cell lines, which was similar to the findings in human lymphocytes. However, in the multidrug-resistant subline the rate of the ouabain-resistant uptake is higher than in the parental cell line, V_{max} being 115 and 78 mmol/(L cells) h and K_m being 0.18 and 0.37 mM, respectively. This difference may be connected to other abnormalities in these cells such as changes in the cytoplasmic pH and in the Na$^+$/H$^+$ antiport of the membrane (see Brismar et al., 1995a).

12.5.4. Muscle

Measurements in frog sartorius muscle demonstrated that the muscle fiber membrane cannot distinguish between Tl$^+$ and K$^+$, as long as [Tl$^+$] is kept low (Mullins and Moore, 1960). In the resting fiber the distribution of Tl$^+$ between fiber water and the bathing solution is similar to that of K$^+$, and changes in the extracellular concentration has the same depolarizing action (58 mV/10-fold increase) for both ionic species. The efflux of Tl$^+$ is increased during stimulation to the same extent (300-fold) as the increase in K$^+$ efflux is stimulated.

In cultured chick myocardial cells some 60% of the Tl$^+$ uptake occurs via an active ouabain-inhibitable mechanism with an apparent K_m of 2–7 μM. Increasing [K$^+$] caused competitive inhibition of the Tl$^+$ influx, where the active component had much higher (260–900 times) affinity for Tl$^+$ than K$^+$ (McCall et al., 1985). Hypoxia has only small or insignificant effects on the Tl$^+$ kinetics in cultured myocardial cells where the rate constant for uptake was 0.16 ± 0.07 and for wash out 0.26 ± 0.06 min^{-1} (Friedman et al., 1987).

An in vivo analysis was performed on the ability of skeletal muscle to accumulate Tl$^+$ (Brismar, 1991). A mixture of ^{201}Tl was injected intramuscularly in the rat, by which ^{99m}Tc-pertechnetate served as a reference substance with negligible intracellular accumulation (Fig. 7). After 30 min 8.9% of the injected ^{99m}Tc-pertechnetate and 49% of ^{201}Tl remained in the muscle, which indicated that the difference (40%) of the ^{201}Tl content was intracellular. The half-time of the intracellular Tl$^+$ accumulation was 4.9 min. Ouabain inhibited the intracellular Tl$^+$ accumulation by 38%. This figure may be an underestimate because the concentration of ouabain is likely to be insufficient in parts of the tissue after the injectate containing ^{201}Tl and ouabain has spread intramuscularly.

12.5.5. Epithelial Cells

An indication of the mechanism of Tl$^+$ uptake in lung epithelial cells is found in the studies of the rate of ^{86}Rb accumulation in the presence or absence of ouabain and furosemide (or bumetanide). ^{86}Rb uptake was twice as fast in freshly isolated tracheal pneumocytes as in alveolar macrophages. In lung epithelial cells 55–60% is ouabain-inhibitable and an additional 15–20% is inhibited by loop diuretics. The rate of ^{86}Rb uptake in fetal cells was <10% of

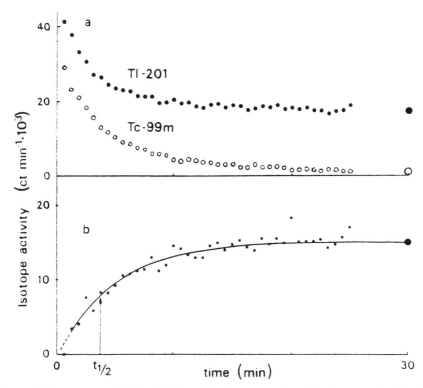

Figure 7. Time course of ^{201}Tl and ^{99m}Tc activity in muscle after single intramuscular injection of isotope mixture (0.1 mL). ^{99m}Tc-pertechnetate is not taken up by the cells and therefore served as a marker for the amount of isotopes remaining in the extracellular compartment (*Upper part*). The difference (after scaling to the same initial value) indicated the intracellular accumulation of [201]Tl (*Lower part*).

the rate in adult cells, and a larger fraction (80–85%) was ouabain-inhibitable in fetal cells (Bland and Boyd, 1986). In alveolar epithelial (type II) cells from rats the rate of K^+ transport measured by ^{86}Rb uptake is accounted for by Na^+,K^+-ATPase activity and $Na^+-K^+-Cl^-$ cotransport amounting to 10.0 and 8.9 nmol/mg protein min, respectively. In guinea pigs, the ouabain-sensitive component is quantitatively similar to that in the rat, whereas the ouabain-insensitive fraction is less [5.1 nmol/(mg protein min)]. Furthermore, in guinea pigs the ouabain-insensitive uptake consists of two approximately equal components, one is the $Na^+-K^+-Cl^-$ cotransport, the other is due to K^+-Cl^- cotransport and K,H-ATPase activity (Kemp et al., 1994).

Epithelial cells in the distal part of the mammalian colon play a critical role in K^+ homeostasis. Del Castillo et al. (1994) studied the rate by which different K^+ transporters contribute to the uptake of [86]Rb. Forty percent is ouabain-sensitive, 29% is blocked by bumetanide, and the remaining 31% was unidentified. They found that one third of the ouabain-sensitive component did not require intracellular Na^+ for its function. Presumably this was due to the presence of a Na^+-independent ouabain-sensitive K^+-ATPase, which has been identified in the apical membrane of colonic cells (Watanabe et al., 1989).

12.6. CONCLUSION

The ouabain-inhibitable $Na^+/K^+(Tl^+)$-ATPase-dependent transport and the furosemide- (or bumetanide-) inhibitable mechanism for cotransport of $Na^+,K^+(Tl^+)$ and Cl^- are the two main routes for Tl^+ uptake, in most cells together accounting for almost 90% of the transport capacity. Both these transporters carry Tl^+ with 5–10 times higher efficiency than K^+. There is a multitude of K^+ channels in the membrane, all showing approximately 20% higher permeability for Tl^+ than K^+, through which Tl^+ may leak from the cells following its electrochemical gradient.

REFERENCES

Adams, D. J., Dwyer, T. M., and Hille, B. (1980). The permeability of endplate channels to monovalent and divalent metal cations. *J. Gen. Physiol.,* **75**, 493–510.

Apell, H.-J. (1989). Electrogenic properties of the Na,K pump. *J. Membr. Biol.,* **110**, 103–114.

Ashcroft, F. M., and Stanfield, P. R. (1983). The influence of the permeant ions thallous and potassium on inward rectification in frog skeletal muscle. *J. Physiol.,* **343**, 407–428.

Bakker-Grunwald, T. (1979). Movement of thallium ion across the ascites cell membrane. *J. Membrane Biol.,* **47**, 171–183.

Bell, J., and Miller, C. (1983). Effects of phospholipid surface charge on ion conduction in the K^+ channel of the sarcoplasmic reticulum. *Fourth Biophysical Discussion* Biophysical Society, Airlie House (Virginia), pp. 223–228.

Bielen, F. V., Glitsch, H. G., and Verdonck, F. (1991). Dependence of Na^+ pump current on external monovalent cations and membrane potential in rabbit cardiac Purkinje cells. *J. Physiol.* (Lond.), **442**, 169–189.

Bielen, F. V., Glitsch, H. G., and Verdonck, F. (1993). Na^+ pump current-voltage relationships of rabbit cardiac purkinje cells in Na^+-free solution. *J. Physiol.,* (Lond.), **465**, 699–784.

Bland, R. D., and Boyd, C. A. R. (1986). Cation transport in lung epithelial cells derived from fetal, newborn and adult rabbits. *J. Appl. Physiol.,* **61**, 507–515.

Blatz, A. L., and Magleby, K. L. (1984). Ion conductance and selectivity of single calcium-activated potassium channels in cultured rat muscle. *J. Gen. Physiol.,* **84**, 1–23.

Blatz, A. L., and Magleby, K. L. (1986). Single apamin-blocked Ca-activated K^+ channels of small conductance in cultured rat skeletal muscle. *Nature,* **323**, 718–720.

Blatz, A. L., and Magleby, K. L. (1987). Calcium-activated potassium channels. *Trends Neurosci.,* **10**, 463–467.

Brismar, T. (1991). In vivo analysis of intracellular thallium-201 accumulation in skeletal muscle of the rat. *Acta. Physiol. Scand.,* **142**, 475–480.

Brismar, T., and Collins, V. P. (1989). Inward rectifying potassium channels in human malignant glioma cells. *Brain Res.,* **480**, 249–258.

Brismar, T., and Collins, V. P. (1993). Effect of external cation concentration and metabolic inhibitors on membrane potential of human glial cells. *J. Physiol.*, (Lond.), **460**, 365–383.

Brismar, T., Collins, V. P., and Kesselberg, M. (1989). Thallium-201 uptake relates to membrane potential and potassium permeability in human glioma cells. *Brain. Res.*, **500**, 30–36.

Brismar, T., Lefvert, A.-K., and Jondal, M. (1994). A method for analysis of cellular K-transport mechanisms through thallium (^{201}Tl) uptake in human lymphocytes. *Acta Physiol. Scand.*, **150**, 299–303.

Brismar, T., Gruber, A., and Peterson, C. (1995a). Increased cation transport in mdrl gene expressing K562 cells. *Cancer Chemother. Pharmacol.*, **36**, 87–90.

Brismar, T., Anderson, S., and Collins, V. P. (1995b). Mechanism of high K^+ and Tl^+ uptake in cultured human glioma cells. *Cell. Mol. Neurobiol.*, **15**, 351–360.

Britten, J. S., and Blank, M. (1968). Thallium activation of the (Na^+–K^+)-activated ATPase of rabbit kidney. *Biochim. Biophys. Acta*, **159**, 160–166.

Chepilko, S., Zhuo, H., Sackin, H., and Palmer, L. G. (1995). Permeation and gating properties of a cloned renal K^+ channel. *Am. J. Physiol.*, **268**, C389–C401.

Chipperfeld, A. (1986). The (Na^+-K^+-Cl^-) co-transport system. *Clin. Sci.*, **71**, 465–476.

Connor, J. A., and Stevens, C. F. (1971). Voltage clamp studies of a transient outward membrane current in gastropod neural somata. *J. Physiol.*, (Lond.), **213**, 21–30.

Del Castillo, J. R., Sulbaran-Carrasco, M. C., and Burgillos, L. (1994). K^+ transport in isolated guinea pig colonocytes: evidence for Na^+-independent ouabain-sensitive K^+ pump. *Am. J. Physiol.*, **266**, G1083–G1089.

Demo, S. D., and Yellen, G. (1992). Ion effects on gating of the Ca^{2+}-activated K^+ channel correlate with occupancy of the pore. *Biophys. J.*, **61**, 639–648.

Eisenman, G., Latorre, R., and Miller, C. (1986). Multi-ion conduction and selectivity in high conductance Ca^{2+}-activated K^+ channel from skeletal muscle. *Biophys. J.*, **50**, 1025–1034.

Eisner, D. A., and Lederer, W. J. (1979). The role of the sodium pump in the effects of potassium-depleted solutions on mammalian cardiac muscle. *J. Physiol.*, **294**, 279–301.

Favari, L., and Mourelle, M. (1985). Thallium replaces potassium in activation of the (Na^+, K^+)-ATPase of rat liver plasma membranes. *J. Appl. Toxicol.*, **5**, 32–34.

Feldman, R. D. (1992). *Beta*-Adrenergic inhibition of Na-K-C-cotransport in lymphocytes. *Am. J. Physiol.*, **263**, C1015–C1020.

Forte, J. G., Ganser, A. L., and Ray, T. K. (1976). The K^+-stimulated ATPase from oxyntic glands of gastric mucosa. In *Gastric Hydrogen Ion Secretion*, Sachs, G. Kasbekar, D. K., and Rehm, W. S. (eds.), Dekker, New York, pp. 302–330.

Fox, J., and Ciani, S. (1985). Experimental and theoretical studies on Tl^+ interactions with the cation-selective channel of the sarcoplasmic reticulum. *J. Membrane Biol.*, **84**, 9–23.

Friedman, B. J., Beihn, R., and Friedman, J. P. (1987). The effect of hypoxia on thallium kinetics in cultured chick myocardial cells. *J. Nucl. Med.*, **28**, 1453–1460.

Geck, P., and Heinz, E. (1986). The Na-K-2Cl cotransport system. *Membrane Biol.*, **91**, 97–105.

Gehring, P. J., and Hammond, P. B. (1964). The interrelationship between thallium and potassium in animals. *J. Pharmacol. Exp. Therapeutics*, **155**, 187–201.

Gorman, A. L. F., Woolum, J. C., and Cornwall, M. C. (1982). Selectivity of the Ca^{2+}-activated and light-dependent K^+ channels for monovalent cations. *Biophys. J.*, **38**, 319–322.

Hagiwara, S., and Takahashi, K. (1974). The anomalous rectification and cation selectivity of the membrane of a starfish egg. *J. Membr. Biol.*, **18**, 61–80.

Hagiwara, S., Eaton, D. C., Stuart, A. E., and Rosenthal, N. P. (1972). Cation selectivity of the resting membrane of squid axon. *J. Membr. Biol.*, **9**, 373–384.

Hagiwara, S., Miyazaki, S., Krasne, S., and Ciani, S. (1977). Anomalous permeabilities of the egg cell membrane of a starfish in K^+-Tl^+ mixtures. *J. Gen. Physiol.*, **70**, 269–281.

Hille, B. (1972). The permeability of the sodium channel to metal cations in myelinated nerve. *J. Gen. Physiol.,* **59,** 637–658.

Hille, B. (1973). Potassium channels in myelinated nerve: Selective permeability to small cations. *J. Gen. Physiol.,* **61,** 669–686.

Hille, B. (1992). *Ionic Channels of Excitable Membranes,* 2nd ed. Sinauer Assoc. Sunderland, Massachusetts.

Hofer, A. M., and Machen, T. E. (1992). K-induced alkalinization in all cell types of rabbit gastric glands: A novel K/H exchange mechanism. *J. Membrane Biol.,* **126,** 245–256.

Jensen, S. K., Jessen, F., and Hoffman, E. K. (1993). Na^+, K^+, Cl^- cotransport and its regulation in Ehrlich ascites tumor cells. Ca^{2+}/calmodulin and protein kinase C dependent pathways. *J. Membrane Biol.,* **131,** 161–178.

Karlish, S. J. D. (1989). The mechanism of cation transport by the Na^+, K^+-ATPase. In *Ion Transport,* Keeling, D., and Benham, C. (eds.). Academic, London, pp. 19–34.

Kemp, P. J., Roberts, G. C., and Boyd, C. A. R. (1994). Identification and properties of pathways for K^+ transport in guinea-pig and rat alveolar epithelial type II cells. *J. Physiol.* (Lond.), **476,** 79–88.

Kimelberg, H. K., and Frangakis, M. V. (1985) Furosemide- and bumetanide-sensitive ion transport and volume control in primary astrocyte cultures from rat brain. *Brain Res.,* **361,** 125–134.

Lytle, C., and Forbush, B., III. (1992). The Na-K-Cl cotransport protein of shark rectal gland. II. Regulation by direct phosphorylation. *J. Biol. Chem.,* **267,** 25438–25443.

McCabe, R. D., and Young, D. B. (1992). Evidence of a K^+-H^+-ATPase in vascular smooth muscle cells. *Am. J. Physiol.,* **262,** H1955–H1958.

McCall, D., Zimmer, L. J., and Katz, A. M. (1985). Kinetics of thallium exchange in cultured rat myocardial cells. *Circ. Res.,* **56,** 370–376.

Mullins, L. J., and Moore, R. D. (1960). The movement of thallium ions in muscle. *J. Gen. Physiol.,* **43,** 759–773.

Omay, H. S., and Schwarz, W. (1992). Voltage-dependent stimulation of Na/K-pump current by external cations: selectivity of different K congeners. *Biochim. Biophys. Acta,* **1104,** 167–173.

Owen, N. E., and Prastein, M. L. (1985). Na/K/Cl cotransport in cultured human fibroblasts. *J. Biol. Chem.,* **260,** 1445–1451.

Post, R. L., Merritt, C. R., Kingsolving, C. R., and Albright, C. D. (1960). Membrane adenosine triphosphatase as a participant in the active transport of sodium and potassium in the human erythrocyte. *J. Biol. Chem.,* **235,** 1796–1802.

Rabon, E. C., and Reuben, M. A. (1990). The mechanism and structure of the gastric H,K ATPase. *Ann. Rev. Physiol.,* **52,** 321–344.

Reichelt, W., and Pannicke, T. (1993). Voltage-dependent K^+ currents in guines pig Müller (glial) cells show different sensitivities to blockade by Ba^{2+}. *Neurosci. Lett.,* **155,** 15–18.

Reuter, H., and Stevens, C. F. (1980). Ion conductance and ion selectivity of potassium channels in snail neurons. *J. Membr. Biol.,* **57,** 103–118.

Robinson, J. D. (1975). Functionally distinct classes of K^+ sites on the (Na^+ + K^+)-dependent ATPase. *Biochim. Biophys. Acta,* **384,** 250–264.

Rogawski, M. A. (1985). The A-current: How obiquitous a feature of excitable cells is it? *Trends Neurosci.,* **8,** 214–219.

Saier, M. H., and Boyden, D. A. (1984). Mechanism, regulation and physiological significance of the loop diuretic-sensitive NaCl/KCl symport system in animal cells. *Mol. Cell. Biochem.,* **59,** 11–32.

Sessler, M. J., Geck, P., Maul, F.-D., Hör, G., and Munz, D. L. (1986). New aspects of cellular thallium uptake: Tl^+-Na^+-$2Cl^-$-cotransport is the central mechanism of ion uptake. *Nucl. Med.,* **25,** 24–27.

Skulskii, I. A., Manninen, V., and Glasunov, V. V. (1990). Thallium and rubidium permeability of human and rat erythrocyte membrane. *Gen. Physiol. Biophys.*, **9**, 39–44.

Skulskii, I. A., Manninen, V., and Järnefelt, J. (1978). Factors affecting the relative magnitudes of the ouabain-sensitive and the ouabain-insensitive fluxes of thallium ion in erythrocytes. *Biochim. Biophys. Acta*, **506**, 233–241.

Stanfield, P. R., Aschcroft, F. M., and Plant, T. D. (1981). Gating of a muscle K^+ channel and its dependence on the permeating ion species. *Nature*, **289**, 509–511.

Stürmer, W., Bühler, R., Apell, H.-J., and Läuger, P. (1991). Charge translocation by the Na,K-pump. II. Ion binding and release at the extracellular face. *J. Membr. Biol.*, **121**, 163–176.

Tas, P. W. L., Massa, P. T., and Koschel, K. (1986). Preliminary characterization of an $Na^+/K^+/Cl^-$ co-transport activity in cultured human astrocytes. *Neurosci. Lett.*, **70**, 369–373.

Tas, P. W. L., Massa, P. T., Kress, H. G., and Koschel, K. (1987). Characterization of an $Na^+/K^+/Cl^-$ co-transport in primary cultures of rat astrocytes. *Biochim. Biophys. Acta*, **903**, 411–416.

Taylor, P. S. (1987). Selectivity and patch measurements of A-current channels in Helix aspersa neurones. *J. Physiol.*, **388**, 437–447.

van Driel, I. R. and Callaghan, J. M. (1995). Proton and potassium transport by H^+/K^+-ATPases. *Clin. Exp. Pharmacol. Physiol.*, **22**, 952–960.

Watanabe, T., Suzuki, T., and Suzuki, Y. (1989). Ouabain-sensitive K ATPase in epithelial cells from guinea pig colon. *Am. J. Physiol.*, **258**, G506–G511.

Wollmuth, L. P., and Hille, B. (1992). Ionic selectivity of I_h channels of rod photoreceptors in tiger salamanders. *J. Gen. Physiol.*, **100**, 749–765.

13

EFFECT OF THALLIUM TOXICITY ON THE VISUAL SYSTEM

Homayoun Tabandeh

Moorfields Eye Hospital, London, England EC1V 2PD

13.1. BACKGROUND

Since its discovery in 1861 by William Crooks, thallium has been responsible for many occupational, accidental, deliberate, and therapeutic poisonings. In the past it had been used in the treatment of syphilis, gonorrhea, dysentery, tuberculosis, and ringworm of the scalp (Crooks, 1861; Saddique and Patterson,

Thallium in the Environment, Edited by Jerome O. Nriagu.
ISBN 0-471-17755-5. © 1998 John Wiley & Sons, Inc.

1983). Poisoning resulted from the systemic administration, absorption from depilatories applied to skin, or inhalation of contaminated dust. Lamy found that in high doses this heavy metal is lethal to animals (Lamy, 1863). Buschke and Langer (1927) described the pathological effects of thallium sulfate. In the first half of this century, soluble acetate and sulfate salts of thallium were widely used as a pesticide. Present use includes the production of semiconductors, scintillation counters, and imitation jewelry. Its ability to transmit long-wave radiation has given thallium a role in lens manufacture. It is also used as a chemical catalyst and in clinical practice in cardiac scanning.

Thallium is colorless, odorless, and tasteless. These properties have made it an ideal poison both for rodents and for use by psychopathic poisoners (Burnett, 1990; Cavanagh, 1991). Its use as a homicidal poison is said to be widespread (Moeschlin, 1988). Numerous reports have discussed the systemic effects and toxicology of thallium poisoning.

The clinical picture of thallium intoxication is varied; the diagnosis is often suspected in the presence of the triad of symptoms:

1. Alopecia and skin rash
2. Painful peripheral neuropathy
3. Confusion and lethargy

The lethal dose for humans is 15–20 mg/kg; nonfatal intoxication occurs below this dose. The first symptoms of poisoning are abdominal pain, gastroenteritis, tachycardia, and headache, which usually occur within 12 h. A characteristic dark, pigmented band appears in the scalp hair within 4 days. Neurological symptoms appear in 2–5 days and include confusion, hallucinations, and convulsions. In severe cases coma, respiratory paralysis, and death occurs within 1 week. When smaller doses are taken, painful peripheral sensory-motor neuropathy and ataxia are the outstanding symptoms. Other neurological features include cranial nerve palsies, optic neuropathy, choreoathetosis, tremor, and encephalopathy. Scalp alopecia is the best known symptom of chronic thallium poisoning, which begins 10 days after the ingestion and is completed within 1 month (Kallner, 1946; Grossman, 1956). Skin may be involved by acneiform eruptions, a papulomacular rash, and dystrophy of the nails (Mee's strips). Autonomic dysfunction, hypertension, cardiomyopathy, electrocardiographic changes, testicular toxicity, hypokalaemia, renal failure, abnormal liver function, leukocytosis, and thrombocytopenia have been reported (Table 1). Electroencephalography (EEG) shows nonspecific slow-wave activity, and electromyography (EMG) is suggestive of distal axonopathy.

Ophthalmologic features of thallium intoxication include effects on the anterior segment of the eye, the retina, optic nerve, and other cranial nerves.

Table 1 Ophthalmological Manifestations of
Thallium Intoxication (Summary)

Loss of lateral half of eyebrows
Lid skin lesions
Ptosis
Seventh nerve palsy
External ophthalmoplegia
Nystagmus
Noninflammatory keratitis
Lens opacities
Mydriasis
Optic neuropathy, optic atrophy
Abnormalities of visual function
    Impairment of contrast sensitivity
    Abnormal color vision, tritanomaly
    Impaired visual acuity
    Central, caecocentral scotomas
    Impairment of ERG
    Delayed visual evoked response (VER)

13.2. ANTERIOR SEGMENT

Ocular tissues appear to store thallium in high concentrations by two possible mechanisms. The first involves binding with melanin in the pigmented tissues such as the iris, ciliary body, and the choroid. The second is by active transport such as that seen in the crystalline lens and the neuroretina (Potts and Au, 1971). High tissue concentrations can result in local toxicity. Animal experiments demonstrated cataract formation, keratitis, and intraocular inflammation following poisoning with thallium. Mice lens has a tendency to accumulate thallium, and administration of thallium to rats leads to the development of cataracts. Dogs poisoned with thallium develop conjunctivitis, keratitis, and corneal ulcers, with histological examination showing inflammatory cellular infiltration (Richet, 1899; Ginsberg and Buschke, 1923; Andre, et al., 1960; von Sallmann et al., 1963). Intravenous administration of thallium to rhesus monkeys results in conjunctivitis, as well as peripheral polyneuropathy and cerebellar ataxia.

Local effect of thallium on animal ocular tissues has also been described. Experiments involving 10-min exposure of the cornea debrided of epithelium to a neutral saturated solution of thallium chloride resulted in corneal oedema, which gradually cleared from the periphery, but left a nebulous central area (Grant and Kern, 1956).

In human thallium poisoning, anterior segment involvement is rare. Mild reversible snowflake cataract has been reported (Kubesova, 1949; Shamshinova et al., 1990); however, conjunctivitis, keratitis, and uveitis are rare features.

13.3. RETINA

Although clinical retinal changes appear to be rare, subclinical and functional impairments have been described. Hennekes reported the electroretinogram (ERG) changes observed in a human case of acute poisoning. Changes included abnormal scotopic β-wave amplitude and impaired amplitude–intensity relation of the β-wave component (Hennekes, 1983). It was suggested that thallium has a direct toxic effect on the retinal tissue. This observation was consistent with the histological findings (Manschot, 1969).

In asymptomatic subjects with low-grade chronic industrial exposure to thallium, Shamshinova and colleagues (1990) reported subnormal ERG particularly in individuals with recent exposure and suggest ERG as a means of testing for early, preclinical thallium toxicity. In the same report electrophysiological studies carried out on rabbits injected with intravitreal doses of thallium demonstrated that the retina, particularly the photoreceptor layer, is susceptible to the effect of thallium, and the degree of impairment depends on the duration and dose of exposure (Shamshinova et al., 1990). It is likely that retinal damage contributes to the optic atrophy, which often accompanies thallium intoxication (Grant and Schuman, 1993).

13.4. OPTIC NEUROPATHY

In 1927 Kaps and Krauss described the development of optic nerve atrophy following thallium ingestion (Kaps, 1927; Krauss, 1927). Optic neuropathy occurs more often in chronic than in acute form of intoxication. In single acute poisoning the incidence of optic nerve damage is about 25%, but in repeated poisoning the optic nerve is involved in the majority of the cases (Bohringer, 1952). Optic neuropathy seen in thallium poisoning is usually retrobulbar, causing blurring of vision, central or cecocentral scotomas, and abnormality of the color vision with a normal ophthalmoscopic appearance of the optic disk in the early stages (Tabandeh and Thompson, 1993; Tabandeh et al., 1994). When the anterior part of the optic nerve is involved, disk swelling accompanies visual disturbance. Optic atrophy is a frequent sequel of optic nerve involvement that can lead to permanent visual loss in repeated poisoning. Lange (1952) reported bilateral optic neuropathy developing 5 weeks following oral administration of two doses of 0.75 g of thallium sulfate, 10 days apart. Optic atrophy ensued weeks later. Optic neuropathy occurring after systemic absorption from a topical preparation were reported in the 1930s. The cases were related to the use of Koremlu cream, a depilatory cream containing 7% thallium acetate (Lillie and Parker, 1932; Mahoney, 1932;

Rudolphy, 1935). Mahoney (1932) described three cases of chronic poisoning with features including failing vision, central scotomas, peripheral polyneuropathy, and hair loss, following regular application of the depilatory cream to the face and arms (Mahoney, 1932). Lillie and Parken (1932) reported pain and paraesthesia developing a month after initial application of the cream, with progressive weakness and muscular wasting over the next 6 months and failing vision after a year.

Contrast sensitivity and color vision are abnormal before the appearance of other clinical signs of optic neuropathy and provide early evidence of optic nerve involvement (Tabandeh and Thompson, 1993; Tabandeh et al., 1994).

13.5. OCULAR NERVE PALSY

Cranial nerves other than the optic nerve may also be involved as part of polyneuropathy observed in thallium intoxication, although their involvement is less common than that of the optic nerve (Pentschew, 1958). Isolated palsies of the fourth and sixth cranial nerves leading to ocular motility defects have been reported (Lange, 1952; Pentschew, 1958). Cavanagh and co-workers (1974) reported degenerative changes in extraocular muscle nerve fibers in a case of external ophthalmoplegia following severe poisoning. Davis and co-workers reported complete external ophthalmoplegia and facial paralysis in a patient with very severe intoxication before any evidence of optic nerve atrophy developed. Other forms of motility impairments, ptosis, and facial nerve palsy may also occur.

13.6. HISTOLOGICAL FINDINGS

Histopathological studies have documented peripheral axonal degeneration as the main pathological change. Davis reported primarily axonal degeneration, with preservation of most of the overlying myelin in the cranial and peripheral nerves of a patient with severe poisoning (Davis et al., 1981). Following intravenous administration of thallium to rhesus monkeys, histological changes included evidence of damage to optic nerves, optic tract, and the lateral geniculate body.

Damage to the neuronal tissue can also involve the retina. Manschot reported the histological changes in a 55-yr-old woman who died 19 yr after severe thallium toxicity, causing alopecia, polyneuritis, psychosis, and optic neuritis leading to optic atrophy. He found a selective loss of retinal ganglion cells, while the outer nuclear layer and the photoreceptors appeared undamaged. There was marked nerve fiber loss. No abnormality was detected in the anterior segments of the eyes (Manschot, 1969).

Thallium retinal toxicity is further supported by animal experiments involving intravitreal injection of thallium. Shamshinova and co-workers (1990) reported selective damage to the photoreceptor layer within 2 days of injection. Although no abnormality in retinal pigment epithelium (RPE) and inner retinal layer was observed on light microscopy, electron microscopy revealed the presence of membrane cytoplasmic inclusions within the RPE cells as well as abnormal orientation of the photoreceptors and shortening of the outer segments.

13.7. COMMENTS

The prognosis of thallotoxicosis depends on the total amount ingested, the time course of the ingestion, and the individual susceptibility. The reported mortality rate is 6–13% with over half of the survivors being left with residual neurological deficit. Abnormal vision together with tremor, muscle weakness, convulsions, and coma are associated with residual neurological deficit (Munch, 1934; Reed et al., 1963). The main mechanism for visual loss is through the involvement of the optic nerve. Bohringer (1952) reported that despite partial optic nerve atrophy the vision often improves for the first year. Lillie and Parker (1932) also reported partial improvement of the vision after discontinuation of thallium-containing depilatory cream. The snowflake cataracts and the anterior chamber inflammation have also been reported to regress (Kubesova, 1949). The persisting residual visual deficit often consists of abnormality of color vision and scotomas.

The mechanism for toxic effects of thallium is not fully understood. A number of pathophysiological processes may be involved. Owing to the similarity of ionic radii, thallium and potassium are exchangeable, resulting in considerable disruptive effect in the cell function. A high affinity for sodium–potassium–ATPase and –SH containing enzymes causes disruption of cell membrane stability, ion flux mechanisms, and the pK. There are also close similarities between thallium intoxication and thiamine deficiency and arsenical neuropathies. Thallium neurotoxicity may be related to the interaction of this ion with riboflavin by forming insoluble complexes and intracellular sequestration of riboflavin, with consequent effects upon energy generation mechanisms associated with tissue flavoproteins (Cavanagh, 1991). Other possible mechanisms include disruption of calcium homeostasis and alterations in amino acids and neurotransmitters (Mulkey and Oehme, 1993). Any of these mechanisms can affect the visual system.

Interference with potassium-dependent processes and respiratory systems, binding with sulfhydryl group of membrane enzymes, alteration of amino acids and neurotransmitters, and impairment of intracellular calcium can all result in functional impairment of the neuronal pathways, retina, and the crystalline lens. Secondary degenerative changes result in permanent deficit in visual function.

To summarize, ophthalmological manifestations of intoxication with thallium include loss of the lateral half of the eyebrows, lid skin lesions, ptosis, seventh nerve palsy, internal and external ophthalmoplegia, and nystagmus. Noninflammatory keratitis and lens opacities have also been described (Richet, 1899; Castelao, 1941; Duke-Elder, 1947), and optic neuropathy is well reported (Kaps, 1927; Lillie and Parker, 1932; Kallner, 1946; Allson, 1953; Symonds, 1953). Functional changes include impairment of contrast sensitivity, abnormal color vision (tritanomaly), impaired visual acuity, and central or caecocentral scotomas (Table 1). There is also abnormality of ERG and a delayed VER. The changes in visual function are essentially those related to the optic nerve and retinal damage.

REFERENCES

Allsop, J. (1953). Thallium poisoning. *Aust. Ann. Med.,* **2,** 144.

Andre, T., Ulberg, S., and Winguist, G. (1960). Accumulation and retention of thallium in tissues of the mouse. *Acta Pharmacol. Toxic.,* **16,** 229–234.

Bohringer, H. R. (1952). Opticusschadigungen durch Thallimvergiftung. *Praxis,* **41,** 1092–1094.

Burnett, J. W. (1990). Thallium poisoning. *Cutis,* **46,** 112–113.

Buschke, A., and Langer, E. (1927). Die Forensische und gewerbl-hygienische Bedeutung des Thalliums. *München med. Wchnschr.,* **74,** 1494–1497.

Castelao, A. M. (1941). Catarata del talio. *Med. Española,* **5,** 395–405.

Cavanagh, J. B. (1991). What have we learnt from Graham Frederick Young? Reflections on the mechanism of Thallium poisoning. *Neuropathol. Appl. Neurobiol.,* **17,** 3–9.

Cavanagh, J. B., Fuller, N. H., Johnson, H. R. M., and Rudge, P. (1974). The effects of thallium salts, with particular reference to to the nervous system changes. *O.J. Med.,* **43,** 293–319.

Crooks, W. (1861). On existence of a new element probably of sulphur group. *Chem. News,* **3,** 193.

Davis L. E., Standefer, J. C., Kornfeld, M., Abercrombie, D. M., and Butler, C. (1981). Acute thallium poisoning: Toxicological and morphological studies of the nervous system. *Ann. Neurol.,* **10,** 38–44.

Duke-Edler, W. S.(1947). *Textbook of Ophthalmology.* London, p. 3153.

Ginsberg, S., and Buschke, A. (1923). Augenveranderungen bei Ratten Nach Thalliumfutterung. *Klin. Mbl. Augenheilk.,* **71,** 385–399.

Grant, W. M., and Kern, H. L. (1956). Cations and the cornea. *Am. J. Ophthalmol.,* **42,** 167–181.

Grant, W. M., and Schuman, J. S. (1993). *Toxicology of the Eye.* Charles C Thomas, Springfield, Illinois.

Grossman, H. (1956). Thallotoxicosis: Report of case and review. *Paediatrics,* **16,** 868–872.

Hennekes, R. (1983). Impairment of retinal function in a case of thallium intoxication. *Klin. Mbl. Augenheilk.,* **182,** 334–336.

Kallner, A. (1946). Thallium-Vergiftung bei Kinder. *Ann. Paediat.,* **167,** 188.

Kaps, L. (1927). Kriminelle tödliche subakute Thallium Vergiftung. *Wien klin. Wchnschr.,* **40,** 967–970.

Krauss (1927). Demonstration: Postneuritischer Atrophie beider Sehnerven. *Klin. Mbl. Augenheilk.,* **79,** 829.

Kubesova, J. (1949). Neuritis retrobulbaris and cataract in thallium intoxication. *Cs. Oftal.,* **5,** 149–153.

Lamy, M. A. (1863). Sur les effets toxiques du thallium. *C.R. Acad. Sci., Paris,* **57,** 442–445.

Lange, F. (1952). Doppleseitige Optikusatrophie nach akuter Thalliumvergiftung. *Klin. Mbl. Augenheilk,* **121,** 221–223.

Lillie, W. I., and Parker, H. L. (1932). Retrobulbar neuritis due to thallium poisoning. *JAMA,* **98,** 1347–1349.

Mahoney, W. (1932). Retrobulbar neuritis due to thallium poisoning from depilatory cream. Report of 3 cases. *JAMA,* **98,** 618–620.

Manschot, W. A. (1969). Ophthalmic pathological findings in a case of thallium poisoning. *Ophthalmolog. Addit. Ad.,* **158,** 348–349.

Moeschlin, S. (1988). Thallium poisoning. *Clin. Toxicol.,* **17,** 133–146.

Mulkey, J. P. and Oehme, F. W. (1993). A review of thallium toxicity. *Vet. Human Toxicol.,* **35,** 445–453.

Munch, J. C. (1934). Human thallotoxicosis. *JAMA,* **102,** 1929–1934.

Pentschew, A. (1958). *Intoxikationen. Handbuch der speziellen pathologischen Anatomie und Histologie. Erkrankungen des zentralen Nervensystems.* Berlin, Springer, Berlin, pp. 1907–2502.

Potts, A. M., and Au, P. C. (1971). Thallous ion and the eye. *Invest. Ophthalmol.,* **10,** 925–931.

Reed, D., Crawley, J., Faro, S., Pieper, S., and Kurland, L. (1963). Thallotoxicosis: Acute manifestations and sequelae. *JAMA,* **187,** 96–102.

Richet, C. (1899). De la toxicite du thallium. *C.R. Soc. Biol. Paris,* **1,** 252–255.

Rudolphy, J. B. (1935). Optic atrophy due to thallium. *Arch. Ophthalmol.,* **13,** 1108–1109.

Saddique, A., and Patterson, C. (1983). Thallium poisoning: A review. *Ve. Hum. Toxicol.,* **25,** 16–26.

Shamshinova, A., Ivanina, A., Yakovlev, L., Shabalina, V., and Spiridonova, V. (1990). Electroretinography, in the diagnosis of thallium intoxication. *J. Hyg. Epidemiol. Microbiol. Immunol.,* **34,** 113–121.

Symonds, W. J. C. (1953). Alopecia, optic atrophy and peripheral neuritis of probably toxic origin. *Lancet,* **II,** 1338–1339.

Tabandeh, H., and Thompson, G. M. (1993). Visual function in thallium toxicity [letter; comment]. *Bmj.,* **307,** 6899.

Tabandeh, H., Crowston, J. G., and Thompson, G. M. (1994). Ophthalmologic features of thallium poisoning. *Am. J. Ophthalmol.,* **117,**(2); 243–245.

von Sallmann, L., Grimes, P., and Collins, E. (1963). Triparanol induced cataract in rats. *Arch. Ophthalmol.,* **70,** 522–529.

INDEX